T0210968

Communications in Computer and Information Science   856

Commenced Publication in 2007
Founding and Former Series Editors:
Alfredo Cuzzocrea, Xiaoyong Du, Orhun Kara, Ting Liu, Dominik Ślęzak,
and Xiaokang Yang

Editorial Board

Simone Diniz Junqueira Barbosa
>   *Pontifical Catholic University of Rio de Janeiro (PUC-Rio),*
>   *Rio de Janeiro, Brazil*

Phoebe Chen
>   *La Trobe University, Melbourne, Australia*

Joaquim Filipe
>   *Polytechnic Institute of Setúbal, Setúbal, Portugal*

Igor Kotenko
>   *St. Petersburg Institute for Informatics and Automation of the Russian*
>   *Academy of Sciences, St. Petersburg, Russia*

Krishna M. Sivalingam
>   *Indian Institute of Technology Madras, Chennai, India*

Takashi Washio
>   *Osaka University, Osaka, Japan*

Junsong Yuan
>   *University at Buffalo, The State University of New York, Buffalo, USA*

Lizhu Zhou
>   *Tsinghua University, Beijing, China*

More information about this series at http://www.springer.com/series/7899

Mojisola Erdt · Aravind Sesagiri Raamkumar
Edie Rasmussen · Yin-Leng Theng (Eds.)

Altmetrics for Research Outputs Measurement and Scholarly Information Management

International Altmetrics Workshop, AROSIM 2018
Singapore, Singapore, January 26, 2018
Revised Selected Papers

Springer

Editors
Mojisola Erdt 🆔
Nanyang Technological University
Singapore
Singapore

Aravind Sesagiri Raamkumar 🆔
Nanyang Technological University
Singapore
Singapore

Edie Rasmussen
University of British Columbia
Vancouver, BC
Canada

Yin-Leng Theng
Nanyang Technological University
Singapore
Singapore

ISSN 1865-0929   ISSN 1865-0937   (electronic)
Communications in Computer and Information Science
ISBN 978-981-13-1052-2   ISBN 978-981-13-1053-9   (eBook)
https://doi.org/10.1007/978-981-13-1053-9

Library of Congress Control Number: 2018947189

© Springer Nature Singapore Pte Ltd. 2018
This work is subject to copyright. All rights are reserved by the Publisher, whether the whole or part of the material is concerned, specifically the rights of translation, reprinting, reuse of illustrations, recitation, broadcasting, reproduction on microfilms or in any other physical way, and transmission or information storage and retrieval, electronic adaptation, computer software, or by similar or dissimilar methodology now known or hereafter developed.
The use of general descriptive names, registered names, trademarks, service marks, etc. in this publication does not imply, even in the absence of a specific statement, that such names are exempt from the relevant protective laws and regulations and therefore free for general use.
The publisher, the authors and the editors are safe to assume that the advice and information in this book are believed to be true and accurate at the date of publication. Neither the publisher nor the authors or the editors give a warranty, express or implied, with respect to the material contained herein or for any errors or omissions that may have been made. The publisher remains neutral with regard to jurisdictional claims in published maps and institutional affiliations.

Printed on acid-free paper

This Springer imprint is published by the registered company Springer Nature Singapore Pte Ltd.
part of Springer Nature
The registered company address is: 152 Beach Road, #21-01/04 Gateway East, Singapore 189721, Singapore

Preface

The workshop on Altmetrics for Research Outputs Measurement and Scholarly Information Management (AROSIM 2018) was held on January 26, 2018, at the Wee Kim Wee School of Communication and Information at the Nanyang Technological University in Singapore. The objective of the workshop was to create a forum to disseminate the latest works on Altmetrics for measuring research impact and scholarly information management, and to investigate how social media-based metrics along with traditional and nontraditional metrics can advance the state of the art in measuring research outputs.

The workshop was attended by about 68 international as well as local participants from Singapore, Malaysia, India, China, Germany, Austria, UK, USA, Kuwait, Australia, Hong Kong, Indonesia, Bangladesh, United Arab Emirates, and Nigeria. Participants comprised research staff, faculty, students, librarians, and publishers. A total of seven papers were accepted for presentation at the workshop (with an acceptance rate of 35%), as well as three posters and an in-house demonstration of a scholarly metrics information system (ARIA). In addition, two keynote lectures and a panel discussion on the future of Altmetrics were held as part of this workshop.

We would like to thank the Program Committee members and the Organizing Committee members for their support with the workshop. We would also like to thank all authors who submitted their works to the workshop, as well as all participants of the workshop for making the workshop a success.

We sincerely hope you enjoy reading the proceedings of the AROSIM 2018 workshop.

April 2018

Mojisola Erdt
Aravind Sesagiri Raamkumar
Edie Rasmussen
Yin-Leng Theng

Acknowledgements

The AROSIM workshop was conducted as a part of our Altmetrics research. This research is supported by the National Research Foundation, Prime Minister's Office, Singapore under its Science of Research, Innovation and Enterprise programme (SRIE Award No. NRF2014-NRF-SRIE001-019).

Organization

Workshop Organizing Committee

Yin-Leng Theng	Centre for Healthy and Sustainable Cities (CHESS), Wee Kim Wee School of Communication and Information (WKWSCI), Nanyang Technological University, Singapore
Edie Rasmussen	University of British Columbia, Vancouver, Canada
Yonggang Wen	School of Computer Science and Engineering, Nanyang Technological University, Singapore
Robert Jäschke	Information School, The University of Sheffield, UK
Isabella Peters	Web Science, ZBW Leibniz Information Center for Economics and Christian Albrechts University Kiel, Germany
Yew Boon Chia	Humanities and Social Sciences Library, Nanyang Technological University, Singapore
Samantha Ang	Lee Wee Nam Library, Nanyang Technological University, Singapore
Mojisola Erdt	Centre for Healthy and Sustainable Cities (CHESS), Wee Kim Wee School of Communication and Information (WKWSCI), Nanyang Technological University, Singapore
Aravind Sesagiri Raamkumar	Centre for Healthy and Sustainable Cities (CHESS), Wee Kim Wee School of Communication and Information (WKWSCI), Nanyang Technological University, Singapore

Workshop Program Committee

Anup Kumar Das	Jawaharlal Nehru University, India
Aparna Basu	South Asian University, India
Ehsan Mohammadi	University of South Carolina, USA
Hamed Alhoori	Northern Illinois University, USA
Joanna Sin	Nanyang Technological University, Singapore
Juan Gorraiz	University of Vienna, Austria
Juan Pablo Alperin	Simon Fraser University, Canada
Judit Bar-Ilan	Bar-Ilan University, Israel
Kazunari Sugiyama	National University of Singapore, Singapore
Kim Holmberg	University of Turku, Finland
Kuang-hua Chen	National Taiwan University, Taiwan
Li Xuemei	York University, Canada
Lutz Bornmann	Max Planck Society, Germany

Michael Khor	Nanyang Technological University, Singapore
Mike Taylor	Digital Science, UK
Mike Thelwall	University of Wolverhampton, UK
Paul Groth	Elsevier Labs, USA
Paul Wouters	CWTS, Leiden University, The Netherlands
Philipp Mayr-Schlegel	GESIS, Leibniz Institute for the Social Sciences, Germany
Rich Ling	Nanyang Technological University, Singapore
Schubert Foo	Nanyang Technological University, Singapore
Shima Moradi	National Research Institute for Science Policy, Iran
Stefanie Haustein	University of Montreal, Canada
Sun Aixin	Nanyang Technological University, Singapore
Sybille Hinze	DZHW GmbH, Germany
Tim Evans	Imperial College London, UK
Victoria Uren	Aston University, UK
Vincent Larivière	Université de Montréal, Canada
Winson Peng	Michigan State University, USA
Xiao Xiaokui	Nanyang Technological University, Singapore
Xin Shuai	Indiana University Bloomington, USA
Ying-Hsang Liu	Charles Sturt University, Australia

Contents

Introduction

Introduction to the Workshop on Altmetrics for Research Outputs Measurement and Scholarly Information Management (AROSIM 2018)

Mojisola Erdt[1]([⊠])(iD), Aravind Sesagiri Raamkumar[1](iD),
Edie Rasmussen[2], and Yin-Leng Theng[1]

[1] Centre for Healthy and Sustainable Cities (CHESS),
Wee Kim Wee School of Communication and Information (WKWSCI),
Nanyang Technological University, Singapore, Singapore
{Mojisola.Erdt,aravind002,TYLTHENG}@ntu.edu.sg
[2] University of British Columbia, Vancouver, Canada
edie.rasmussen@ubc.ca

Abstract. "Altmetrics" refers to novel metrics, often based on social media, for measuring the impact of diverse scholarly objects such as research papers, source code, or datasets. Altmetrics complement traditional metrics, such as bibliometrics, by offering novel data points to give a more holistic understanding of the outreach of research outputs. This increased awareness could help researchers and organizations to better gauge their performance and understand the outreach and impact of their research. Altmetrics is related to diverse research areas such as informetrics, social network analysis, scholarly communication, digital libraries, and information retrieval. This workshop on Altmetrics for Research Outputs Measurement and Scholarly Information Management (AROSIM 2018) aims to give a multi-disciplinary perspective on this new and cutting-edge research area by addressing a range of opportunities, emerging tools and methods, along with other discussion topics in the fields of altmetrics, traditional metrics, and related areas.

1 Introduction

Dissemination of research outputs is an important activity in the scholarly research lifecycle. Consequently, tracking the impact of research outputs on new media sources has gained focus in the last few years. On the one hand, researchers increasingly need advanced techniques to track the impact of their research outputs as new forms of scholarly communication emerge. On the other hand, organizations such as universities, libraries, research centers, public and private grant funding agencies, and publishers need systems to synthesize, visualize and analyze scholarly outputs at varying granularity for performance management and decision making. Traditionally, metrics such as citation count, impact factor, and h-index have been employed for this purpose. However, with the advent of social media, researchers themselves can now propagate and share their research outputs with diverse communities, thereby initiating debates

© Springer Nature Singapore Pte Ltd. 2018
M. Erdt et al. (Eds.): AROSIM 2018, CCIS 856, pp. 3–8, 2018.
https://doi.org/10.1007/978-981-13-1053-9_1

and discussions about their research. Altmetrics offer new insights and novel approaches to measure and track the impact of scholarly outputs to an even wider, non-academic audience [1]. Altmetrics research seeks to develop innovative metrics to measure research impact on social media and has been gaining momentum over the last few years.

This workshop on Altmetrics for Research Outputs Measurement and Scholarly Information Management (AROSIM 2018) aims to address opportunities, emerging tools, and methods in the fields of altmetrics, non-traditional metrics, and related areas. The theme of the workshop is "The best of both worlds: cross-metric exploration and validation of Altmetrics and Bibliometrics". The main objective of the AROSIM 2018 workshop is to create a forum to disseminate the latest works on Altmetrics for measuring research impact and scholarly information management. The workshop will investigate how social media-based metrics along with traditional metrics can advance the state-of-the-art in measuring research outputs.

The goals of the workshop are:

(i)   To promote the exchange of ideas and encourage potential collaborations amongst scholars from computer science and information science disciplines, as well as with librarians and industry.
(ii) To investigate challenges and explore solutions for cross-metric exploration and validation, while considering disciplinary differences related to measuring research outputs.
(iii) To showcase innovative altmetric tools, methods, and datasets.
(iv) To provide a discussion platform for the academic communities, librarians, policy makers, and publishers, as well as grant funding agencies.

2   Workshop Overview

A total of 20 papers were reviewed. Seven papers were accepted as full papers and included in the proceedings, while three papers were presented as posters during the workshop. The workshop featured two keynote lectures, two sessions for the presentation of full papers, a poster and interactive demo session, and a panel discussion session.

2.1   Keynote Sessions

The first invited lecture was by Mike Thelwall (University of Wolverhampton, UK) on the topic "Using Altmetrics to Support Research Evaluation" [2]. Professor Thelwall gave perspectives on the rationale, advantages and disadvantages of using alternative indicators (altmetrics) in research impact measurement. He showcased studies on alternative indicators such as (i) correlation with citations, (ii) examples of early impact indicators, (iii) social impact, (iv) health impact, (v) arts & humanities impact, (vi) commercial impact, and (vii) education impact. The lecture ended with a call for future work on testing new indicators, testing their value for different countries/languages, and evaluating the pragmatics of using alternative indicators in real-world applications.

The second invited lecture "Towards Greater Context for Altmetrics" [3] was given by Stacy Konkiel (Altmetric.com). She gave both a contemporary and futuristic overview of Altmetrics and its related contexts. She began with an introduction to the objectives of science, humanities and scientometrics. Thereafter, she focused on three sub-topics (i) the current context of altmetrics, (ii) the required context for altmetrics, and (iii) the actions required from industry and research personnel. The lecture ended with a call for future research on the motivations behind researchers' sharing preferences on social media, the effect of language in altmetrics studies, and the values embedded in organizations such as universities, and institutes that use altmetrics to measure research impact.

2.2 Full Paper Presentation Session 1

The first full paper session comprised of four full paper presentations. The first paper "Monitoring the broad impact of the journal publication output on country level: A case study for Austria" by Gorraiz et al. [4] described a study conducted in Austria to monitor the impact of journal articles from different disciplines. The study aimed to identify the most active disciplines by different metrics, and to test whether there are correlations between the metrics. The study highlights the increase in usage of Twitter by the scholarly community in Austria.

The second paper "New Dialog, New Services with Altmetrics: Lingnan University Library Experience" by Cheung et al. [5], outlined the initiative taken by Lingnan University in Hong Kong to integrate altmetrics within its institutional repository (IR). As an initial case study, the university has integrated full-text download usage metrics in their IR. PlumX data was used as a part of this initiative. The university's long-term vision of integrating other altmetrics in the context of their IR system is also highlighted in the paper. Based on the reviews received, this paper was granted the AROSIM 2018 best paper award.

The third paper "How do Scholars Evaluate and Promote Research Outputs? An NTU Case Study" by Zheng et al. [6] described the findings of a case study conducted with 18 researchers from Nanyang Technological University (NTU) in Singapore, regarding altmetrics and the evaluation of research outputs. Specifically, the study aimed to identify how NTU researchers promote and evaluate their research outputs. Findings indicate that only half of the participants used social networking sites for promoting their research and most researchers still use traditional metrics such as citation counts and the journal impact factor for evaluating their research outputs.

The fourth paper "Scientific vs. Public Attention: A comparison of Top Cited Papers in WoS and Top Papers by Altmetric Score" by Banshal et al. [7], described a bibliometric analysis study performed with the top 100 papers with the highest Altmetric Attention Scores in 2016. In this study, papers that obtained a large number of online mentions are characterized with associated properties such as the publication venue (journal), discipline and author distribution. A complementary analysis of the top 100 most cited papers from Web of Science is also conducted.

2.3   Full Paper Presentation Session 2

The second full paper presentation session, held in the afternoon, comprised of three full paper presentations. The first paper "Field-weighting readership: how does it compare to field-weighting citations?" by Huggett et al. [8], described a study comparing Mendeley readership with citations. The findings highlighted a strong correlation between Mendeley reads and citation counts at an overall general level. The correlation was however found to be lower at a field-weighted level. The authors claimed that field-weighted readership metrics could be useful as a complementary metric to field-weighted citations.

The second paper "A Comparative Investigation on Citation Counts and Altmetrics between Papers Authored by Universities and Companies in the Research Field of Artificial Intelligence" by Luo et al. [9], described a study on bibliometric and altmetric analysis of computer science and artificial intelligence publications. A comparison was performed between publications from academia and industry. The findings from this study indicated that university-authored papers received more citations than industry-authored papers, whereas, in terms of altmetrics, industry-authored papers performed better.

The third paper "Scientometric Analysis of Research Performance of African Countries in selected subjects within the field of Science and Technology" by Utieyineshola [10], was about a study applying scientometric analysis to investigate the research performance of African countries in selected subjects within the field of science & technology, in the last 20 years. Findings from this study indicated that South Africa was the best performing country with the highest number of publications and highest h-index. The research performance of different regions of Africa were also compared.

2.4   Poster and Interactive Demo Session

Three posters were presented during the workshop's lunch break. The first poster "Database-Centric Guidelines for Building a Scholarly Metrics Information System: A Case Study with ARIA Prototype", by Sesagiri Raamkumar, Luo, Erdt and Theng, outlined the design considerations involved in building a database for storing scholarly metrics such as citation counts, publication counts, and tweet counts. Technical issues and their solutions were also presented. The second poster "Evolution and state-of-the art of Altmetric research: Insights from network analysis and altmetric analysis", by Lathabai, Prabhakaran and Changat, aimed to identify the evolutionary trajectories of the altmetrics research field. Key papers that have contributed to this research field were identified in the study. The third poster "Feature Analysis of Research Metrics Systems", by Nagarajan, Erdt and Theng, presented the results of a feature analysis study conducted on 18 research metric systems. A total of 48 individual features from 14 broad categories were used for the analysis. The results of this study give insights into features that could be improved and developed in future.

An interactive demo session was conducted showcasing the ARIA (Altmetrics for Research Impact and Actuation) [11] prototype system. ARIA is a tool developed to help in measuring research impact for researchers, universities, research institutes and policy-makers. The key features of the tool include researcher and institutional dashboards, visualizations of both bibliometric and altmetric indicators, and a cross-metric explorer.

2.5 Panel Discussion Session

A panel discussion on the topic "Altmetrics in the year 2030" was conducted during the workshop. This session was moderated by Michael Khor (Nanyang Technological University). The panel speakers included Schubert Foo (Nanyang Technological University), Na Jin Cheon (Nanyang Technological University), Mike Thelwall (University of Wolverhampton), Stacy Konkiel (Altmetric.com), Edie Rasmussen (University of British Columbia), Tina Moir (Elsevier), and Stephanie Faulkner (Elsevier).

Questions addressed to the panel during the session included:

  (i)   What changes can be expected in Altmetrics in the year 2030, assuming that the existing services continue to be in operation?
 (ii)   What sort of "disruptive" innovations can be expected by 2030?
(iii)   What current gaps/issues in Altmetrics will be rectified by 2030?
 (iv)   What will be the effects of manipulations and other malpractices related to Altmetrics?
  (v)   What will be the impact of Open Access (OA) journals in this area? What sort of changes can be foreseen in the usage of Altmetrics by the year 2030?

3 Outlook

With this first-of-a-kind workshop conducted in the Asia Pacific region, we hope to have created and also sustained interest in Altmetrics as a young and exciting research field. It can be observed that the community is still growing. Our intention through this workshop has been to create a sustainable bridge between altmetrics, bibliometrics, scientometrics, research evaluation and other closely related fields. The success of a research area is dependent for its sustenance and improvement not just on the increasing number of publications every year but also on the participation of different stakeholders. With this workshop, we were successful in bringing together academicians, librarians, publishers, and research managers. We sincerely hope that the future research directions proposed by the speakers in this workshop will be actively pursued by the research community.

Acknowledgements. This research is supported by the National Research Foundation, Prime Minister's Office, Singapore under its Science of Research, Innovation and Enterprise programme (SRIE Award No. NRF2014-NRF-SRIE001-019).

We would also like to thank Altmetric.com for sponsoring the AROSIM 2018 best paper award.

References

1. Priem, J., Taraborelli, D., Groth, P., Neylon, C.: Altmetrics: a manifesto (2010)
2. Thelwall, M.: Using altmetrics to support research evaluation. In: Erdt, M., Sesagiri Raamkumar, A., Rasmussen, E., Theng, Y.-L. (eds.) AROSIM 2018. CCIS, vol. 856, pp. 11–28. Springer, Singapore (2018)
3. Konkiel, S.: Towards greater context for altmetrics. In: Erdt, M., Sesagiri Raamkumar, A., Rasmussen, E., Theng, Y.-L. (eds.) AROSIM 2018. CCIS, vol. 856, pp. 29–35. Springer, Singapore (2018)
4. Gorraiz, J., Blahous, B., Wieland, M.: Monitoring the broad impact of the journal publication output on country level: a case study for Austria. In: Erdt, M., Sesagiri Raamkumar, A., Rasmussen, E., Theng, Y.-L. (eds.) AROSIM 2018. CCIS, vol. 856, pp. 39–62. Springer, Singapore (2018)
5. Cheung, S.L., Kot, C.F., Chan, Y.K.: New dialog, new services with altmetrics: Lingnan University library experience. In: Erdt, M., Sesagiri Raamkumar, A., Rasmussen, E., Theng, Y.-L. (eds.) AROSIM 2018. CCIS, vol. 856, pp. 63–71. Springer, Singapore (2018)
6. Zheng, H., Erdt, M., Theng, Y.L.: How do scholars evaluate and promote research outputs? An NTU case study. In: Erdt, M., Sesagiri Raamkumar, A., Rasmussen, E., Theng, Y.-L. (eds.) AROSIM 2018. CCIS, vol. 856, pp. 72–80. Springer, Singapore (2018)
7. Banshal, S.K., Basu, A., Singh, V.K., Muhuri, P.K.: Scientific vs. public attention: a comparison of top cited papers in WoS and top papers by Altmetric Score. In: Erdt, M., Sesagiri Raamkumar, A., Rasmussen, E., Theng, Y.-L. (eds.) AROSIM 2018. CCIS, vol. 856, pp. 81–95. Springer, Singapore (2018)
8. Huggett, S., Palmaro, E., James, C.: Field-weighting readership: how does it compare to field-weighting citations? In: Erdt, M., Sesagiri Raamkumar, A., Rasmussen, E., Theng, Y.-L. (eds.) AROSIM 2018. CCIS, vol. 856, pp. 96–104. Springer, Singapore (2018)
9. Luo, F., Zheng, H., Erdt, M., Sesagiri Raamkumar, A., Theng, Y.L.: A comparative investigation on citation counts and altmetrics between papers authored by universities and companies in the research field of artificial intelligence. In: Erdt, M., Sesagiri Raamkumar, A., Rasmussen, E., Theng, Y.-L. (eds.) AROSIM 2018. CCIS, vol. 856, pp. 105–114. Springer, Singapore (2018)
10. Utieyineshola, Y.: Scientometric analysis of research performance of african countries in selected subjects within the field of science and technology. In: Erdt, M., Sesagiri Raamkumar, A., Rasmussen, E., Theng, Y.-L. (eds.) AROSIM 2018. CCIS, vol. 856, pp. 115–124. Springer, Singapore (2018)
11. Nagarajan, A., Sesagiri Raamkumar, A., Luo, F., Vijayakumar, H., Erdt, M., Theng, Y.L.: Altmetrics for Research Impact Actuation (ARIA): a multidisciplinary role-based tool for cross-metric validation. In: 4: AM Altmetrics Conference, Toronto, Canada (2017)

Keynote Papers

Using Altmetrics to Support Research Evaluation

Mike Thelwall$^{(\boxtimes)}$ (iD)

University of Wolverhampton, Wolverhampton, UK
m.thelwall@wlv.ac.uk

Abstract. Altmetrics are indicators that have been proposed as alternatives to citation counts for academic publication impact assessment. Altmetrics may be valued for their speed or ability to reflect the non-scholarly or societal impacts of research. Evidence supports these claims for some altmetrics but many are limited in coverage (the proportion of outputs that have non-zero values) or ability to reflect societal impact. This article describes data sources for altmetrics, indicator formulae, and strategies for applying them for different tasks. It encompasses traditional altmetrics as well webometric and usage indicators.

Keywords: Altmetrics · Webometrics · Research evaluation

1 Introduction

Altmetrics were first narrowly defined as indicators for academic activities or outputs derived from the social web. This definition was conceived by PhD student Jason Priem and promoted by him and collaborators primarily through a highly successful October 2010 manifesto [1]. The goals included identifying non-academic impacts of scholarly publications and generating more up-to-date indicators than possible with citation counts. The development of altmetrics had become possible in 2010 not just through the rise of the social web but also by the free provision of automated access to data from many social web services. Thus, for example, interest in an article on Twitter could be tracked in real time using the free Twitter Applications Programming Interface (API) to find citations or links to articles. The manifesto also argued that altmetrics should not be taken at face value but that research was needed to help interpret their value and meanings. Altmetrics have subsequently developed to the extent that information professionals can be expected to know the basic facts about them [2]. The UK LIS Bibliometrics Forum has developed a set of guidelines for the competencies needed for people conducting quantitative research evaluations of various types (https://thebibliomagician.wordpress.com/competencies/). This separates the knowledge and skills needed into three levels: entry, core and specialist. Some knowledge of altmetrics is required here at entry level.

The altmetrics initiative was not the first to propose the use of online indicators for faster and wider impact evidence for scholarly outputs. Web indicators had previously been created by the field of webometrics (e.g., [3]). The main difference between altmetrics and webometrics is that web-based indicators are more difficult to collect and

© Springer Nature Singapore Pte Ltd. 2018
M. Erdt et al. (Eds.): AROSIM 2018, CCIS 856, pp. 11–28, 2018.
https://doi.org/10.1007/978-981-13-1053-9_2

typically rely upon web crawling or automated queries in commercial search engines to identify online citations to publications, scholars, or organisations [4]. For example, citations from Wikipedia to academic outputs could be obtained by crawling Wikipedia (or accessing a copy of Wikipedia's content) and extracting citations. Alternatively, citations could be identified by using automated Bing queries to search for mentions of a set of academic outputs in Wikipedia [5]. This is possible with a site-specific query that combines the first author last name, publication title and year, as in the following example for a Bulletin of the American Mathematical Society article that searches only the Malaysian version of Wikipedia.

- Alexanderson 2006 "Euler and Königsberg's bridges A historical view" site:my. wikipedia.org

The meaning of the term altmetrics has since expanded to encompass all digital traces of the use or impact of academic outputs. Both ImpactStory and Altmetric.com include some webometric indicators within their altmetric portfolio. Thus, altmetrics no longer need to originate from the social web but can also be webometrics (from the general web) and usage indicators, such as download counts. The term altmetrics now connotes any alternative indicator of academic impact, where citation counts are the standard against which altmetrics are the alternative.

This article explains the theoretical background of altmetrics and describes reasons to use them, data sources, more complex indicators derived from raw altmetric data, and research into their meaning. The focus is on altmetrics for academic articles rather than altmetrics for books, data, software or other scholarly outputs (for a review, see: [6]). This article includes altmetrics from social media platforms that are popular in the USA, such as Twitter, because most research so far covers these. Evidence is needed to test whether data from other sites, such as Weibo, vk.com and FC2, has similar properties. Altmetrics are heterogenous in value, type of impact reflected (if any), and useful applications [7] and so each one should be considered on its own merits.

2 Background: Citations and Altmetrics for Research Evaluation

Academic research needs to be evaluated on many different levels to support decision making. As some of the examples below suggest, at any given level evaluators may compare their unit with comparable other units to benchmark their performance or may analyse their sub-units to support internal decisions.

- At the international level, newspapers and others may evaluate university research to generate league tables to promote themselves. Scientometricians may analyse science to gain theoretical insights.
- At the national level, government policy makers may analyse the performance of a country's researchers to assess their international competitiveness and scientific strengths to inform future research funding or other policy decisions. They may analyse national performance trends over time to assess the impact of an important

policy change (e.g., restructuring or funding). They may compare the country's universities to allocate research funding or inform policy decisions.

- At the sub-national level, research funders' evaluation teams may analyse the performance of the research funded by them to assess the effectiveness of their funding strategy or individual funding streams.
- At the university level, bibliometricians may compare the performance of different departments to inform decisions about how to share internal funding or whether to close ineffective units. They may compare the university against other universities to predict future shares of performance-based research income.
- At the department or research group level, research administrators may compare the performance of a group against international competitors to make the case for the success of its research, or may evaluate the research to identify individual successful researchers or research directions.
- At the individual manager level, leaders may evaluate individual researchers for promotion, tenure or hiring decisions.
- At the individual researcher level, scientists may evaluate their own research for feedback on their own performance, to promote their achievements or to identify successful publications.

The above list is not exhaustive because local circumstances can generate other applications. For example, UK departments need to identify the best four publications of each researcher within a specified set of years for periodic national research assessments [8], creating a specific need for publication-level selection assessments within departments. Academic journals are also often evaluated and compared [9]. The scale and purpose of an analysis affect whether to employ altmetrics or citation counts.

Research is traditionally evaluated using peer review. When articles are submitted to academic journals or monographs are submitted to scholarly presses, they are typically reviewed by subject specialists to decide whether they can be published. Expert judgement is also the main technique used to evaluate research for some of the purposes listed above. For example, a manager may read job applicants' papers and CVs to decide on the research strengths of the candidates. At the opposite extreme, a government may recruit hundreds of experts to read the main research outputs and associated statements of its academics to decide how much funding each department should receive [8].

Expert judgements are sometimes informed by citation count data or altmetrics (e.g., [10]), to cross-check the outcomes for bias, to provide a tie-breaker when reviewers disagree, or to provide an initial overall judgement to be critiqued. The indicator data can help speed up the review process, saving valuable (and expensive) reviewer time, and providing an alternative perspective to allow reviewers to cross-check their judgements and perhaps re-evaluate problematic cases.

Citation counts or altmetrics are sometimes used without substantial expert review. This may occur when the cost of the amount of expert review needed would be greater than the benefits of the greater accuracy for decision making. For example, an academic wondering whether their first paper was better received than their second paper might compare their citation counts rather than recruiting three experts to read them both and report on their relative merits. At the other scale extreme, if the Chinese government wanted to know whether the average quality of its research had increased over the past ten years, then it might not wish to pay experts to read and evaluate a large enough random sample from each year to reach a conclusion. It might instead commission a citation analysis, accepting that the results may be imperfect.

2.1 Citation Analysis

Citation-based indicators are widely used to support research evaluation at all levels. The underlying reason why citation counts can reflect scholarly impact is that the references in a paper to some extent list the prior works that influenced it [11]. A paper that has been referenced (i.e., cited) by many different publications has therefore apparently been very influential. On this basis, comparing the citation counts of papers or groups of papers may reveal which has been most influential. This argument is imperfect for many reasons, however, including the following. Solutions are suggested after each point in brackets.

1. Some citations are negative, perfunctory or arbitrary [12]. In these cases, the cited work has not influenced the citing work. This seems to be common in the arts and humanities. [Action: accept that citation counts are sometimes misleading and avoid drawing strong conclusions in the arts and humanities.]
2. Decisions about which papers to cite may be influenced by biasing factors, such as author nationality, prestige, university and collaborations (e.g., [13, 14]). [Action: accept that citation counts can include systematic biases and take steps to check for these or acknowledge this limitation; consider using appropriate academic-related altmetrics in addition.]
3. Research can be useful in academic in ways that do not generate citations, such as by closing off false lines of research or giving information to academics in ways that it is not conventional to cite [15]. [Action: accept that low cited work may be valuable in academia.]
4. Average citation counts vary by document type, field and year [16] and so it is not fair to compare citation counts between papers from different fields and years. [Action: report citation counts alongside the average number of citations per article for the same document type, field and year, or calculate a normalised citation count indicator for documents of the same type that factors out the field and year of publication.]
5. Citations may not reflect valuable non-academic uses of research and so may not be good indicators of societal impact [17]. [Action: consider using appropriate altmetrics in addition.]
6. It takes about three years for citation counts to be mature enough to be a reasonable indicator of the long-term impact of a paper (a three-year citation window is common, e.g., [18]). This delay is unacceptable for the many research evaluations that focus on, or cannot ignore, recent research. [Action: consider using social web altmetrics in addition.]

In summary, citation counts can help research evaluations if employed carefully and if reported with enough context to be interpreted appropriately by decision makers.

2.2 Altmetrics

Altmetrics partly address three of the above limitations of citation counts: time delays and inability to reflect some academic and non-academic impacts [1, 19, 20]. They can address the time delay problem by collecting data from the social web, which is typically much faster to appear than a citation. For example, someone might read an article in the week in which it is published and then tweet about it immediately and register it in their Mendeley library. If they then used that article to influence their research, it might be several years before that research was completed, written up, refereed and published in early view so that Scopus or the Web of Science (WoS) could index its citations.

Altmetrics partly address the problem of tracing non-citation impacts because, in theory, an article might be mentioned in the social web by anybody that uses it. A member of the public might tweet about an article that had given them insights into their medical condition or a social worker might blog about an article that had influenced their decision about a difficult case. Nevertheless, it seems likely that when articles are found useful, there is rarely a digital record of this. Altmetrics can only therefore hope to track a fraction of the impacts of academic research.

An extra advantage is that altmetrics can reveal where research has had an impact, which may inform future dissemination strategies [21].

Altmetrics do not address some problems of citation analysis [22, 23]. An altmetric mention of an article may be negative or perfunctory, and decisions about which article to mention in the social web seem to be more likely to be influenced by biasing social factors, including nationality and friendship. National citation biases are likely to be stronger in social web altmetrics due to higher uptake in the USA for many services [24]. From a wider impact perspective, most uses of academic research probably do not leave digital traces that can be identified through altmetrics [25] and so they can only reflect a fraction of an article's impact.

Altmetrics have one additional problem that citations do not: they are relatively easy to manipulate. Thus, for example, it is possible to generate thousands of tweets about an article and an academic might do this if they believed that their promotion or funding would depend on it. Thus, altmetrics need to be used much more cautiously than citations and should not be used in evaluations when those evaluated are aware of their use in advance and would benefit from a favourable outcome – there would be too much temptation to cheat [26].

Three initiatives have surveyed and reported on aspects of altmetrics. Snowball Metrics (https://www.snowballmetrics.com/) produced a set of recommendations for university self-evaluations. The National Information Standards Organization (NISO) in the USA launched an initiative to promote standards and best practice for altmetrics during 2013–2016. Its final report discusses data standards, gives use cases for how altmetrics might be exploited and has a section on altmetrics for research data [27]. The European Commission Expert Group on Altmetrics focused on the relationship between altmetrics and open science, in the sense of scholarship that that is accessible to the public [28]. It made recommendations about how altmetrics could support open science.

3 Data Sources: Observation, Purchase and Harvesting

There are three main ways for individual scholars and research managers to access altmetric data: observation (e.g., noticing or finding altmetric scores for a research output in a publisher's website); bulk purchasing scores from an altmetric data provider; and using software to harvest scores directly from the sources.

3.1 Observation

Many publisher websites now report one or more altmetrics within journal or article pages. Thus, when accessing an article online, the number of times it has been tweeted, cited or blogged may be visible within its page. For example, each PLOS article has a "Metrics" tab listing various citation count, download and altmetric data. Similarly, Wiley-Blackwell articles may have an Altmetric button that links to a page of Altmetric.com scores and ScienceDirect papers (Elsevier) may list data from PlumX at the side. These indicators are presumably for the article's potential readers and authors but may also be used by publishers and editors to monitor their journals.

The altmetric data provided by some publishers may have been collected by them (e.g., download and view counts) or may be imported from scholarly data providers, such as Altmetric.com, Plum Analytics (part of Elsevier from February 2017), and ImpactStory. These specialist organisations continually harvest altmetric data from the social web using the APIs of sites like Twitter as well as web crawling and other methods. They then package sets of indicators for academics or publishers.

ImpactStory, for example, creates altmetric profiles for researchers by combining evidence for all their publications. At the time of writing, their example profile was for Ethan White (https://profiles.impactstory.org/u/0000-0001-6728-7745), reporting that he had 4470 online mentions in the previous year from Twitter, Facebook, G+, Reddit and other sources. The profile page also lists the online mentions for each of the academic's individual publications. Altmetric.com also offers a free browser-based "bookmarklet" (https://www.altmetric.com/products/free-tools/bookmarklet/). Users installing it can see altmetric scores for articles accessed in their web browser.

3.2 Purchase

Universities, research councils and governments can bulk purchase altmetric data from the commercial altmetric providers Altmetric.com (https://www.altmetric.com/products/explorer-for-institutions/) and Plum Analytics (https://plumanalytics.com/products/plumx-metrics/). A university might pay for counts of altmetric mentions of all its publications together with a range of analytic tools to identify strengths and weaknesses and to compare impact between departments. These analytics may also benchmark the university or individual departments against world norms.

Altmetric.com also provides its raw data free for researchers (http://api.altmetric.com/), giving a simple source for scientometricians conducting altmetric research (see: [29]).

3.3 Harvesting

Some altmetric scores can be automatically collected free from the source using software, bypassing the need to pay to access the data. This can be time consuming and the software can take time to master, making it impractical for casual users but a good solution for scientometricians. For altmetric research, the ability to harvest scores directly from the source allows researchers to have full control over what data is collected and how it is collected. Thus, as part of a research design it might be important that all the scores are collected at approximately the same time, which may not be the case if they are downloaded from data providers.

To illustrate automatic data collection, the free Windows software Webometric Analyst (http://lexiurl.wlv.ac.uk) supports altmetric data gathering from Mendeley and Twitter as well as many different types of webometric data gathering. Webometric Analyst can be fed with a list of publications and it will then query the Mendeley API to find out the number of times that each one has been registered by a Mendeley user. This gives the Mendeley reader count altmetric score. As discussed below in the evidence section, this is a good early indicator of the likely future impact of an article. Webometric Analyst can also be fed with a list of DOIs or article URLs and then it will continually query Twitter for mentions of them. The Twitter API only reports tweets from the previous week and so to count the number of tweets for an article, its DOI or URL must be monitored continually from the moment that it is first available online. Webometric Analyst can do this if it is left running on a computer that is never switched off. Thus, it is more difficult to collect good Twitter data than good Mendeley data. Webometric Analyst can also collect the following webometric scores for any set of journal articles: Wikipedia citations; syllabus mentions; grey literature mentions; general web mentions; web patent citations; PowerPoint citations.

There do not seem to be free programs to collect other altmetric scores (e.g., from Facebook, Google Plus, Weibo). If data is needed from these and Altmertric.com free research data cannot be used then a program would need to be written to collect it.

4   Research Evaluation Indicators

Altmetric data consists of a simple count of the number of times that a research output has been cited, mentioned or downloaded. When altmetrics are used for research evaluations it is necessary to compare these counts between publications, or groups of publications – or to assess whether the numbers are high or low. A simple and effective way to do this is through percentiles: calculating the percentage of other outputs that have higher (or lower) scores. Thus, Altmetric.com might report that an article has a "High Attention Score compared to outputs of the same age and source (92nd percentile)" (https://www.altmetric.com/details/20710370) and ImpactStory might reveal that "Your top publication has been saved and shared 111 times. Only 9% of researchers get this much attention on a publication" (https://profiles.impactstory.org/u/0000-0002-0072-029X).

For formal evaluations, it is important to ensure fairness so that no researchers or groups are disadvantaged through unreasonable comparisons. The most important three sources of bias are document type, publication year and field of study.

- **Document type:** Review articles are more likely to be cited than standard articles, despite not reporting original research. Thus, review articles are normally excluded or treated separately in research evaluations. It would be unfair to compare books or conference papers with journal articles since they are likely to have different citation rates. There are also contributions to journals that are less likely to be cited, such as editorials, letters and notes. These should be excluded or treated separately.
- **Field:** Citation rates vary between fields because the structure of the field, typical reference list lengths, and referencing norms affect the type and number of documents that are cited. It is not clear that these issues transfer to altmetric data although there are known to be major field differences in average altmetric scores, so it is important to take field into account for altmetrics.
- **Publication year:** Other factors being equal, older documents tend to be more cited than younger documents because they have had more time to be referenced. Time is also a factor for altmetrics but older documents may have lower scores on altmetrics because the main interest in them occurred before the social web.

The simplest fair solution to these issues is only to compare an altmetric score with the scores of other outputs of the same type, field and year. Thus, with the percentile approach it would be reasonable to report the percentage of documents of the same type that had lower scores within the same field and year. A practical disadvantage is that the scores for all other documents of the same type, field and year are needed (or a random sample) to calculate the percentiles but this is unavoidable if a score is to be interpreted against the world average. Altmetric data providers can calculate percentiles easily because they attempt to gather comprehensive data sets but this can be time consuming or impractical for individual researchers. Currently (October 2017), public altmetric scores from the main providers normalise for publication year but not field of study.

4.1 Field and Year Normalised Average Citations/Mentions Per Publication

If groups of publications are to be compared against each other using altmetrics then it would be natural to compute the average altmetric score for the set and then compare the averages for different groups against each other (e.g., to compare journals: [9]) or with the world average. To be fair, these comparisons should only be performed between sets of documents of the same field, publication year and type for the reasons given above. This is often impractical because groups of researchers tend to publish in multiple fields and research evaluations often include outputs from several years.

It is possible to address the above problem by using an averaging method that takes into account the field and year of publication. This solution is a field (and year) normalised indicator of average impact, such as the Mean Normalised Citation Score (MNCS) [30] or the Mean Normalised Log Citation Score (MNLCS) variant [31]. These work by first dividing each score by the world average for the field and year and then averaging the normalised values. In this way, the averages of the normalised scores can be compared between years and fields. In theory, the formulae could also normalise for publication type but in practice they are usually applied exclusively to journal articles (often excluding reviews, editorials, notes, etc.).

Irrespective of the number of publication years and fields, an MNCS or MNLCS calculation always results in a single average for the whole group. This average can be interpreted as follows.

- MNCS or MNLCS score of 1: The group's outputs, on average, tend to score the same as the rest of the world's outputs from the corresponding fields and years.
- MNCS or MNLCS score > 1: The group's outputs, on average, tend to score *higher* than rest of the world's outputs from the corresponding fields and years.
- MNCS or MNLCS score < 1: The group's outputs, on average, tend to score *lower* than rest of the world's outputs from the corresponding fields and years.

The difference between MNCS and MNLCS is that scores are log transformed for MNLCS to reduce the influence of very high values. This is important because otherwise the results can be dominated by individual outliers. Using the MNLCS variant of MNCS therefore makes the results more precise [31].

4.2 Field and Year Normalised Proportion Cited/Mentioned

The MNCS and MNLCS average score methods for groups of articles do not work well if most of the article scores are zero. This is common for most altmetrics, with Mendeley readers and Twitter citations being the main exceptions. This problem can be resolved by reporting instead the *proportion* of a group's outputs with a nonzero score. Comparing proportions between groups would be unfair because of field, year and publication type differences but the solution is to use an indicator that normalises for theses. The Equalised Mean-based Normalised Proportion Cited (EMNPC) solves this problem in a similar way to the MNCS and MNLCS by dividing the proportion for each field and year by the world average proportion cited for that field and year, with an additional calculation to reduce bias [31]. As for MNLCS, values above 1 are above the world average and values below 1 are below it.

5 Empirical Evidence

Empirical evidence has been gathered about the coverage of altmetrics (the proportion of outputs that have a non-zero score) and the correlation between altmetric scores and citation counts for different fields and years. Some studies have focused on a single altmetric whereas others have assessed a range. A few studies have also attempted to assess the type of impact that individual altmetrics reflect.

5.1 Coverage

The coverage (proportion of outputs with a non-zero score) of an altmetric affects the purposes for which it can be used. If most outputs have a score of zero then the altmetric is not discriminatory enough to rank (or compare) individual outputs. If an overwhelming majority of altmetric scores are zero then academics could not be ranked or compared based on their publications. Large groups of researchers could still be compared based upon the proportion cited (EMNPC), however. The coverage of any altmetric will naturally vary by field and year and may also change over time but the following discussion illustrates broad current patterns.

- **High coverage:** The altmetric with the highest coverage is Mendeley reader counts (41–81% from Altmetric.com data: [32]; see also: [33]). At the end of the publication year, the average Mendeley reader count for journal articles is above 1 for most fields, suggesting that most articles have readers [34]. Mendeley has higher coverage than other social reference managers (e.g., Bibsonomy: [35]). Virtually all articles can be expected to have been downloaded, so download and view counts have very high coverage. For comparison purposes, two years after publication, WoS and Scopus citations probably have high coverage for most fields and years.
- **Moderate coverage:** Twitter citations seem to occur for 4–17% of articles (based on altmetric data: [32, 36]). WoS and Scopus citations probably have moderate coverage in the publication year in most fields.
- **Low coverage:** Facebook wall posts, blog citations can be expected to occur for under 1–2% of journal articles (estimates based on Altmetric data: [37]). Google Plus citations, media mentions, Reddit citations, and LinkedIn citations can be expected to occur for under 0.1% of journal articles (estimates based on Altmetric data: [37]). Low rates can also be expected for Wikipedia citations (5%: [5]), PowerPoint citations (estimated under 1%: [38]), SlideShare citations (1%–14%: [39]), patent citations (up to 10% but probably 0% in most fields: [40]), and syllabus mentions (0.1%–51%: [38]) tend to have low coverage for journal articles. Up to 10% of articles in high profile journals (e.g., Science, Nature) may be blogged about [41] but the percentage is likely to be much lower for typical journals. Citations from drug and clinical guidelines [42, 43] are likely to be very rare, even for medical articles.

5.2   Correlation with Citations or Expert Judgements

A concern raised by critics of altmetrics is that scores are meaningless: useless articles may have high scores and excellent articles can have low scores. Tweeting bots [44], for example, undermine tweet counts as reflections of human judgements. The main strategy to counteract this argument and present the case that altmetrics are meaningful has been to assess the extent to which altmetrics correlate with citation counts or expert judgement. Statistically significant evidence of a positive correlation in would show that altmetric scores tend to associate with scholarly impact or value and are therefore not meaningless [45].

The correlation with citation counts strategy, whilst giving empirical evidence of scholarly value, partly goes against one of the original goals of altmetrics [1], to provide evidence of non-scholarly impact. Nevertheless, it seems reasonable to believe that most types of impact for scholarly research would associate in some way with citation counts. For example, articles that have a big impact on a profession (e.g., social work, nursing) and get many social media mentions as a result, may also be cited because of their professional impact. They may also be followed up by future studies that cite them because of their success within the profession. In the other influence direction, if articles are assumed to have different underlying quality (e.g., robustness of argument and background, generality of evidence) then better articles seem more likely to have more impact of all types.

Based on the above arguments it seems reasonable to use correlations with citation counts or peer judgements of value as evidence that altmetrics are valuable impact indicators, but not of the type of impact that they reflect. It is difficult to interpret the strength of the correlations found, however, because weak correlations could be due to low numbers [46] or the altmetric reflecting a type of impact that is only loosely related to scholarly impact. Nevertheless, high correlations suggest that an alternative indicator closely reflects academic impact. Evidence of positive correlations is available for many altmetrics.

- **High positive correlations:** Mendeley reader counts have consistently been shown to have moderate or high correlations with citation counts for most fields and years (typical values in the range 0.5–0.8, depending on field and year: [34, 47]). They have at least moderate correlations with expert judgements for quality-filtered articles from the same broad field and year [48]. Download and view data can be expected to have high correlations with citation counts after a few years, although there are large disciplinary differences [49].
- **Low positive correlations:** Tweets [44], webometrics. Facebook wall posts, blog citations, Google+ citations, media mentions, Reddit citations, and LinkedIn citations (Altmetric data: [36, 37]) all have positive correlations in the range 0.1–0.3. From webometrics, Wikipedia citations [5], PowerPoint citations [38], SlideShare citations [39], patent citations [40], drug and clinical guidelines citations [42, 43] and syllabus mentions [38] tend to have low positive correlations with citation counts.
- **No correlation:** Some of the above indicators have correlations that are zero in practice, such as LinkedIn citations, even though they are statistically significantly positive on huge data sets [37].

5.3 Impact Type

The type of impact, if any, indicated by an altmetric is sometimes obvious but can be difficult to detect. Content analysis, interviews and questionnaires have been used to assess impact type for the non-obvious cases.

- **Narrow impact:** It seems reasonable to assume that the type of impact reflected by the following is determined by the nature of the source of the citations: *patent citations* (commercial); *syllabus mentions* (educational) and *media mentions* (public interest). High correlations between Mendeley reader counts and citation counts combined with evidence that Mendeley reader counts indicate article readers [50] and academics and researchers dominating the self-declared users of Mendeley [51] gives strong evidence that *Mendeley reader counts* reflect scholarly impact, albeit with an element of educational impact in some fields [52]. *PowerPoint citations* probably reflect a combination of scholarly and educational impact since both educators and scholars post presentations online. Citations from drug and clinical guidelines [42, 43] have health and medical impacts.

- **Broad impact:** A content analysis of tweets citing academic articles suggest that they rarely indicate specific uses other than general interest [53]. Another study found that most tweeters of a set of academic articles were academics [54]. *Twitter* citations therefore probably tend to reflect attention or academic attention. *Facebook, Google Plus, Weibo* and other general social network sites and microblogs allow similar short posts to Twitter and have a general userbase so probably also reflect attention or academic attention. It seems reasonable to characterise *Wikipedia citations* as informational impact by the nature of this encyclopedia. Within individual fields, the impact type might be narrower and could be described in different terms. For example, Wikipedia citations to health articles may reflect public health information dissemination. Blog citations seem to reflect both public and scholarly interest since articles are often selected for a blog by academics for their public interest value [55].
- **Unknown or varied impact:** *Download and view counts* are perhaps the broadest possible indicator since almost every type of use of an article starts with the article being accessed in some way (paper or online). It is possible that in many fields scholarly uses would dominate others and so download counts would then be scholarly impact indicators (e.g., [56, 57]). The type of impact of *grey literature citations* [58] and *SlideShare citations* [39] probably varies by field. They may reflect business impact in marketing, for example.

5.4   Early Impact

There is considerable evidence that many altmetrics appear before citations, as might be expected. This is clearest for Mendeley reader counts (and downloads: [59]) but it is probably true for most altmetrics except patent and Wikipedia citations. The early appearance of altmetrics allows them to be used as impact indicators before there are enough citations to do this [60]. There are two dimensions to this advantage.

- **Statistical power:** Indicators that are more common than citations may be more powerful in their ability to differentiate between articles or groups of articles. In other words, if group A has had more impact than group B in a given period, there may not be enough citations to suggest this with any confidence but it might be possible with altmetrics if they appear earlier and in sufficient numbers.
- **Within year time bias:** Articles published early in a year have had longer to be noticed than later published articles and so have an unfair advantage when comparisons are made between articles published in the same year. After a few years, this difference might not matter much for citations, especially for fields in which citation rates peak early [61]. For altmetrics that appear earlier than citations, their rate may also peak before citations, and so comparisons between articles for these altmetrics will be less biased by publication month. Thus, a second early impact advantage of altmetrics is for reducing bias when comparing articles published early and late in a year. This is not relevant for comparisons of time-balanced sets of

publications (i.e., a similar spread of publication months), although it may still reduce statistical power in this case. The within year time bias issue is most relevant for comparisons between individual publications or small sets of publications, when time balance may not be achieved.

Mendeley reader counts appear before citations in most or all fields [34]. Overall, article readers appear about a year before article citations. Combined with high coverage and evidence of high correlations with citation counts, this gives confidence that Mendeley reader counts are suitable early impact indicators as a substitute for citation counts. If care is taken with publication dates, Mendeley reader counts can even be used within the year of publication for articles [62, 63]. Early Mendeley reader counts have a strong positive correlation with later Scopus citation counts in many fields, which also supports their use for early impact [64].

Tweets also appear before citation counts. A study of an online medical journal found that 60% of tweets from the first 60 days occurred in the first two days after publication [65]. Early Tweets also correlated positively with later citation counts for this journal [65], although it may be a special case as an online medical journal. Nevertheless, it is not clear that tweet counts reflect impact in many fields and so they have a much weaker case than Mendeley.

Blog citations have been shown to associate with increased future citations [41], although few articles get blogged about.

6 Summary: Responsibly Using Altmetrics

Altmetrics are now part of the infrastructure of digital publishing. They are used by individuals and organisations for self-evaluations and there is some pressure to use them in formal research evaluations. It is important to ensure that altmetrics are employed appropriately and effectively and so this article concludes by reviewing responsible use guidelines.

For citation analysis, the ten principles of the *Leiden Manifesto* [66] are the main guidelines for formal research evaluations. The first, and most important, is, "Quantitative evaluation should support qualitative, expert assessment". As argued above, the extent to which this is possible varies with the scale and purpose of the evaluation. The principles also argue that an evaluation should be sensitive to the local context, field differences, and the purpose of an evaluation.

Two of the Leiden principles can be a problem for altmetrics, "Keep data collection and analytical processes open, transparent and simple" and "Allow those evaluated to verify data and analysis" [66]. If altmetric scores are purchased from data providers that do not give enough information about how the data was collected then the data is not transparent. Similarly, one of the main altmetric indicators, Mendeley reader counts, is inherently not transparent because the number of readers cannot be traced back to the identity of those readers. Thus, if a researcher is told that their paper has the second most readers in a field, there would be no way for them to check that the readers of the most read paper in their field were genuine. Altmetric.com supports transparency by revealing the source of each altmetric score, when possible. If it reports a tweet count of

8 for a paper then clicking links in the site will allow the eight tweets to be read and checked. It separates Mendeley reader counts from indicators that can be verified at source.

The following principles reflect the Leiden Manifesto, taken from the Metric Tide report [8] and adopted by the *UK Forum for Responsible Metrics*.

- Robustness – in terms of accuracy and scope
- Humility – quantitative evaluation should support, but not supplant, qualitative, expert assessment
- Transparency – that those being evaluated can test and verify the results
- Diversity – accounting for variation by research field, and using a range of indicators to reflect and support a plurality of research and researcher career paths across the system
- Reflexivity – recognising and anticipating the systemic and potential effects of indicators, and updating them in response

In summary, there is general agreement that altmetrics have the potential to be useful for some research evaluations but that care must be taken to use them responsibly. Published recommendations so far have been general rather than giving a recipe for different tasks. This is partly due to the many different altmetrics available, the introduction of new sources of altmetrics and technical changes. It is also partly the result of insufficient academic research to give concrete evidence of best practice for applications. This is an important direction for future research.

References

1. Priem, J., Taraborelli, D., Groth, P., Neylon, C.: Altmetrics: a manifesto (2010). http://altmetrics.org/manifesto/
2. Holmberg, K.: Altmetrics for Information Professionals Past, Present and Future. Chandos, Oxford (2015)
3. Vaughan, L., Hysen, K.: Relationship between links to journal Web sites and impact factors. Aslib Proc. **54**(6), 356–361 (2002)
4. Thelwall, M., Kousha, K.: Web indicators for research evaluation, Part 1: citations and links to academic articles from the web. El Profesional de la Información **24**(5), 587–606 (2015). https://doi.org/10.3145/epi.2015.sep.08
5. Kousha, K., Thelwall, M.: Are Wikipedia citations important evidence of the impact of scholarly articles and books? J. Assoc. Inf. Sci. Technol. **68**(3), 762–779 (2017). https://doi.org/10.1002/asi.23694
6. Kousha, K., Thelwall, M.: Web indicators for research evaluation, Part 3: books and non-standard outputs. El Profesional de la Información **24**(6), 724–736 (2015)
7. Sugimoto, C.R., Work, S., Larivière, V., Haustein, S.: Scholarly use of social media and altmetrics: a review of the literature. J. Assoc. Inf. Sci. Technol. **68**(9), 2037–2062 (2017)
8. Wilsdon, J., Allen, L., Belfiore, E., Campbell, P., Curry, S.: The Metric Tide: Independent Review of the Role of Metrics in Research Assessment and Management. HEFCE, London (2015)
9. Haustein, S., Siebenlist, T.: Applying social bookmarking data to evaluate journal usage. J. Informetr. **5**, 446–457 (2011)

10. Piwowar, H., Priem, J.: The power of altmetrics on a CV. Bull. Assoc. Inf. Sci. Technol. **39**(4), 10–13 (2013)
11. Merton, R.K.: The Sociology of Science: Theoretical and Empirical Investigations. University of Chicago Press, Chicago (1973)
12. Krampen, G., Becker, R., Wahner, U., Montada, L.: On the validity of citation counting in science evaluation: content analyses of references and citations in psychological publications. Scientometrics **71**(2), 191–202 (2007)
13. Campbell, F.M.: National bias: a comparison of citation practices by health professionals. Bull. Med. Libr. Assoc. **78**(4), 376 (1990)
14. Pasterkamp, G., Rotmans, J., de Kleijn, D., Borst, C.: Citation frequency: a biased measure of research impact significantly influenced by the geographical origin of research articles. Scientometrics **70**(1), 153–165 (2007)
15. Seglen, P.O.: Citation rates and journal impact factors are not suitable for evaluation of research. Acta Orthopaedica Scandinavica **69**(3), 224–229 (1998)
16. Althouse, B.M., West, J.D., Bergstrom, C.T., Bergstrom, T.: Differences in impact factor across fields and over time. J. Assoc. Inf. Sci. Technol. **60**(1), 27–34 (2009)
17. van Driel, M.L., Maier, M., Maeseneer, J.D.: Measuring the impact of family medicine research: scientific citations or societal impact? Fam. Pract. **24**(5), 401–402 (2007)
18. Glänzel, W., Schubert, A.: A new classification scheme of science fields and subfields designed for scientometric evaluation purposes. Scientometrics **56**(3), 357–367 (2003)
19. Dinsmore, A., Allen, L., Dolby, K.: Alternative perspectives on impact: the potential of ALMs and altmetrics to inform funders about research impact. PLoS Biol. **12**(11), e1002003 (2014)
20. Fenner, M.: What can article-level metrics do for you? PLoS Biol. **11**(10), e1001687 (2013)
21. Maggio, L.A., Meyer, H.S., Artino, A.R.: Beyond citation rates: a real-time impact analysis of health professions education research using altmetrics. Acad. Med. **92**(10), 1449–1455 (2017)
22. Colquhoun, D., Plested, A.: Why you should ignore altmetrics and other bibliometric nightmares (2014). http://www.dcscience.net/2014/01/16/why-you-should-ignore-altmetrics-and-other-bibliometric-nightmares/
23. Livas, C., Delli, K.: Looking beyond traditional metrics in orthodontics: an altmetric study on the most discussed articles on the web. Eur. J. Orthod. (2017). https://doi.org/10.1093/ejo/cjx050
24. Mas-Bleda, A., Thelwall, M.: Can alternative indicators overcome language biases in citation counts? A comparison of Spanish and UK research. Scientometrics **109**(3), 2007–2030 (2016)
25. Ravenscroft, J., Liakata, M., Clare, A., Duma, D.: Measuring scientific impact beyond academia: an assessment of existing impact metrics and proposed improvements. PLoS ONE **12**(3), e0173152 (2017)
26. Wouters, P., Costas, R.: Users, narcissism and control: tracking the impact of scholarly publications in the 21st century. In: Science and Technology Indicators 2012 (STI 2012), pp. 847–857. SURF Foundation, Utrecht (2012)
27. NISO: Outputs of the NISO Alternative Assessment Metrics Project (2016). http://www.niso.org/apps/group_public/download.php/17091/NISO%20RP-25-2016%20Outputs%20of%20the%20NISO%20Alternative%20Assessment%20Project.pdf
28. Wilsdon, J., Bar-Ilan, J., Frodeman, R., Lex, E., Peters, I., Wouters, P.: Next-generation metrics: responsible metrics and evaluation for open science (2017). https://ec.europa.eu/research/openscience/index.cfm?pg=altmetrics_eg

29. Robinson-García, N., Torres-Salinas, D., Zahedi, Z., Costas, R.: New data, new possibilities: exploring the insides of Altmetric.com. El Profesional de La Información **23**(4), 359–366 (2014)
30. Waltman, L., van Eck, N.J., van Leeuwen, T.N., Visser, M.S., van Raan, A.F.: Towards a new crown indicator: an empirical analysis. Scientometrics **87**(3), 467–481 (2011)
31. Thelwall, M.: Three practical field normalised alternative indicator formulae for research evaluation. J. Informetr. **11**(1), 128–151 (2017). https://doi.org/10.1016/j.joi.2016.12.002
32. Haustein, S., Larivière, V., Thelwall, M., Amyot, D., Peters, I.: Tweets vs. Mendeley readers: how do these two social media metrics differ? IT Inf. Technol. **56**(5), 207–215 (2014)
33. Erdt, M., Nagarajan, A., Sin, S.C.J., Theng, Y.L.: Altmetrics: an analysis of the state-of-the-art in measuring research impact on social media. Scientometrics **109**(2), 1117–1166 (2016)
34. Thelwall, M., Sud, P.: Mendeley readership counts: an investigation of temporal and disciplinary differences. J. Assoc. Inf. Sci. Technol. **57**(6), 3036–3050 (2016). https://doi.org/10.1002/asi.2355
35. Borrego, A., Fry, J.: Measuring researchers' use of scholarly information through social bookmarking data: a case study of BibSonomy. J. Inf. Sci. **38**(3), 297–308 (2012)
36. Costas, R., Zahedi, Z., Wouters, P.: Do "altmetrics" correlate with citations? Extensive comparison of altmetric indicators with citations from a multidisciplinary perspective. J. Assoc. Inf. Sci. Technol. **66**(10), 2003–2019 (2015)
37. Thelwall, M., Haustein, S., Larivière, V., Sugimoto, C.: Do altmetrics work? Twitter and ten other candidates. PLoS ONE **8**(5), e64841 (2013). https://doi.org/10.1371/journal.pone.0064841
38. Thelwall, M., Kousha, K.: Online presentations as a source of scientific impact? An analysis of PowerPoint files citing academic journals. J. Am. Soc. Inf. Sci. Technol. **59**(5), 805–815 (2008)
39. Thelwall, M., Kousha, K.: SlideShare presentations, citations, users and trends: a professional site with academic and educational uses. J. Assoc. Inf. Sci. Technol. **68**(8), 1989–2003 (2017)
40. Kousha, K., Thelwall, M.: Patent citation analysis with Google. J. Assoc. Inf. Sci. Technol. **68**(1), 48–61 (2017)
41. Shema, H., Bar-Ilan, J., Thelwall, M.: Do blog citations correlate with a higher number of future citations? Research blogs as a potential source for alternative metrics. J. Am. Soc. Inf. Sci. Technol. **65**(5), 1018–1027 (2014)
42. Thelwall, M., Kousha, K., Abdoli, M.: Is medical research informing professional practice more highly cited? Evidence from AHFS DI Essentials in Drugs.com. Scientometrics **112**(1), 509–527 (2017)
43. Thelwall, M., Maflahi, N.: Guideline references and academic citations as evidence of the clinical value of health research. J. Assoc. Inf. Sci. Technol. **67**(4), 960–966 (2016)
44. Haustein, S., Bowman, T.D., Holmberg, K., Tsou, A., Sugimoto, C.R., Larivière, V.: Tweets as impact indicators: examining the implications of automated "bot" accounts on Twitter. J. Assoc. Inf. Sci. Technol. **67**(1), 232–238 (2016)
45. Sud, P., Thelwall, M.: Evaluating altmetrics. Scientometrics **98**(2), 1131–1143 (2014)
46. Thelwall, M.: Interpreting correlations between citation counts and other indicators. Scientometrics **108**(1), 337–347 (2016). https://doi.org/10.1007/s11192-016-1973-7
47. Thelwall, M.: Are Mendeley reader counts useful impact indicators in all fields? Scientometrics **113**(3), 1721–1731 (2017). https://doi.org/10.1007/s11192-017-2557-x

48. HEFCE: The Metric Tide: Correlation Analysis of REF2014 Scores and Metrics (Supplementary Report II to the Independent Review of the Role of Metrics in Research Assessment and Management) (2015). http://www.hefce.ac.uk/pubs/rereports/Year/2015/metrictide/Title,104463,en.html

49. Halevi, G., Moed, H.F.: Usage patterns of scientific journals and their relationship with citations. In: Context Counts: Pathways to Master Big and Little Data, pp. 241–251 (2014)

50. Mohammadi, E., Thelwall, M., Kousha, K.: Can Mendeley bookmarks reflect readership? A survey of user motivations. J. Assoc. Inf. Sci. Technol. **67**(5), 1198–1209 (2016). https://doi.org/10.1002/asi.23477

51. Mohammadi, E., Thelwall, M., Haustein, S., Larivière, V.: Who reads research articles? An altmetrics analysis of Mendeley user categories. J. Assoc. Inf. Sci. Technol. **66**(9), 1832–1846 (2015)

52. Thelwall, M.: Why do papers have many Mendeley readers but few Scopus-indexed citations and vice versa? J. Librariansh. Inf. Sci. **49**(2), 144–151 (2017). https://doi.org/10.1177/0961000615594867

53. Thelwall, M., Tsou, A., Weingart, S., Holmberg, K., Haustein, S.: Tweeting links to academic articles. Cybermetrics **17**(1) (2013). http://cybermetrics.cindoc.csic.es/articles/v17i1p1.html

54. Tsou, A., Bowman, T.D., Ghazinejad, A., Sugimoto, C.R.: Who tweets about science? In: Proceedings of ISSI 2015 - 15th International Conference of the International Society for Scientometrics and Informetricspp, pp. 95–100. Boğaziçi University Printhouse, Istanbul (2015)

55. Shema, H., Bar-Ilan, J., Thelwall, M.: How is research blogged? A content analysis approach. J. Assoc. Inf. Sci. Technol. **66**(6), 1136–1149 (2015). https://doi.org/10.1002/asi.23239

56. Schloegl, C., Gorraiz, J.: Comparison of citation and usage indicators: the case of oncology journals. Scientometrics **82**(3), 567–580 (2010)

57. Moed, H.F.: Statistical relationships between downloads and citations at the level of individual documents within a single journal. J. Assoc. Inf. Sci. Technol. **56**(10), 1088–1097 (2005)

58. Wilkinson, D., Sud, P., Thelwall, M.: Substance without citation: evaluating the online impact of grey literature. Scientometrics **98**(2), 797–806 (2014)

59. Moed, H.F., Halevi, G.: On full text download and citation distributions in scientific-scholarly journals. J. Assoc. Inf. Sci. Technol. **67**(2), 412–431 (2016)

60. Kudlow, P., Cockerill, M., Toccalino, D., Dziadyk, D.B., Rutledge, A., Shachak, A., Eysenbach, G.: Online distribution channel increases article usage on Mendeley: a randomized controlled trial. Scientometrics **112**(3), 1537–1556 (2017)

61. Larivière, V., Archambault, É., Gingras, Y.: Long-term variations in the aging of scientific literature: from exponential growth to steady-state science (1900–2004). J. Assoc. Inf. Sci. Technol. **59**(2), 288–296 (2008)

62. Maflahi, N., Thelwall, M.: How quickly do publications get read? The evolution of Mendeley reader counts for new articles. J. Assoc. Inf. Sci. Technol. **69**(1), 158–167 (2018). https://doi.org/10.1002/asi.23909

63. Thelwall, M.: Are Mendeley reader counts high enough for research evaluations when articles are published? Aslib J. Inf. Manag. **69**(2), 174–183 (2017). https://doi.org/10.1108/AJIM-01-2017-0028

64. Thelwall, M.: Early Mendeley readers correlate with later citation counts. Scientometrics (in press). https://doi.org/10.1007/s11192-018-2715-9
65. Eysenbach, G.: Can tweets predict citations? Metrics of social impact based on Twitter and correlation with traditional metrics of scientific impact. J. Med. Internet Res. **13**(4), e123 (2011)
66. Hicks, D., Wouters, P., Waltman, L., De Rijcke, S., Rafols, I.: The Leiden Manifesto for research metrics. Nature **520**(7548), 429–431 (2015)
67. Kousha, K., Thelwall, M.: Assessing the impact of disciplinary research on teaching: an automatic analysis of online syllabuses. J. Am. Soc. Inform. Sci. Technol. **59**(13), 2060–2069 (2008)

Towards Greater Context for Altmetrics

Stacy Konkiel$^{(\boxtimes)}$ (iD)

Altmetric LLP, London, UK
stacy@altmetric.com

Abstract. This paper argues for the need for scientometricians to borrow from qualitative analytical approaches to improve the field of altmetrics research. Existing approaches to contextualizing metrics using quantitative means, by researchers and altmetrics data providers alike, are discussed and critiqued. I propose the need for research based in ways of thinking from the humanities and social sciences to understand human motivations for sharing research online to truly understand what altmetrics mean and how they should be interpreted. I explore several ways that the research community and data providers alike can help those who use altmetrics in evaluation move from "data" to "wisdom".

1 Introduction

In the field of research evaluation, the need for context is often implicit. We, scientometricians, usually assume that those who practice evaluation - the department chairs, vice provosts for research, grant reviewers, and so on, will understand that there is no magic bullet, no special number, that can quickly and legibly summarize the influence of a single paper, nor the legacy of a scholar's entire body of work. We trust that they know that impact evidence especially metrics needs to be contextualized in order to be properly interpreted.

Sadly, we are sometimes proven wrong. Metrics can and are used improperly in research evaluation scenarios. From using a Journal Impact Factor to judge the quality of a paper at a glance when low on time, to determining a faculty member's worthiness for tenure based on grant dollars won, decision-makers have been known to misapply metrics in the moments that can count the most.

In this paper, I argue that no single number, nor even a group of numbers, can in isolation help one understand the impact of research at any level. But I do not wish to "throw the baby out with the metrics bathwater" [1] and insist that peer review can ever be the only way we understand the value of a body of work. To do so would be to ignore the environment of scarcity so prevalent in academia that it prevents evaluators from allowing the time and resources to engage fully with research, instead pushing them towards relying upon numeric indicators alone to understand a work's scholarly influence.

Instead, I argue for the need to take a holistic and realistic approach to understanding research impact, one that combines state-of-the-art quantitative approaches with tried and true qualitative and theory-based means of judging research's value. Put simply, to make research evaluation practices more robust and scalable, we need to marry a scientific worldview with a humanistic one and employ modes of thinking

© Springer Nature Singapore Pte Ltd. 2018
M. Erdt et al. (Eds.): AROSIM 2018, CCIS 856, pp. 29–35, 2018.
https://doi.org/10.1007/978-981-13-1053-9_3

from both disciplines. In particular, I promote this approach in the nascent field of altmetrics, which looks to data from the social web to understand the attention paid to, and in some cases the impact of, research.

2 Altmetrics and the Context We Have

In the few short years since the founding of altmetrics as a field, a great deal has already been accomplished to better contextualize the altmetrics data used in research evaluation.

The altmetrics community is comprised of both researchers who study the data and altmetrics aggregation services that gather said data. Each plays an important part in bettering the field to improve research evaluation for all.

Most altmetrics research uses quantitative approaches borrowed from the field of scientometrics to deepen our knowledge of the many variables that can affect the altmetrics that research receives. Recent studies have examined the effects of author nationality on the attention that research receives on regional and international social media sites [2–5]; the effects of author gender upon the attention that research receives on Twitter [6] and in public policy and mainstream media [7]; and the Open Access status of research upon a variety of altmetrics [6, 8], among other factors.

Perhaps the most important context the field has gained from research so far is the understanding of the effects of self-promotion on the altmetrics and usage statistics for research [9]. That's because the perceived threat of gaming upon the trustworthiness of altmetrics data; it is one of the most common concerns that academics have when discussing the limitations of altmetrics. The Erdt study, when taken hand-in-hand with the Adie framework (Fig. 1) for understanding the nuances of self-promotion and gaming [10], make it clear that the "threat" of gaming is not so clear-cut.

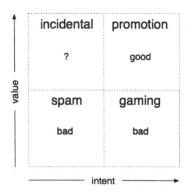

Fig. 1. Adie framework

Altmetrics aggregation services have also worked to contextualize the altmetrics they deliver to researchers and evaluators alike. Altmetric offers numerical percentiles for the Altmetric Attention Score, putting that weighted score in context by comparing it with other research published in the same time period and journal. Impactstory offers "Achievement" badges for authors for attention received in particular online sources (e.g. an author's overall body of work's citedness in Wikipedia warrants a "Wikitastic" badge with contextualized metrics included, see Fig. 2).

Wikitastic Top 10%
Your research is mentioned in 7
Wikipedia articles! Only 6% of
researchers are this highly cited in
Wikipedia.

Fig. 2. Impactstory "Achievement" badge for "Wikitastic" research

But even with this heightened sense of nuance for the metrics we produce and consume, we're still mostly looking at numbers. We need even greater context for the field.

3 Altmetrics and the Context We Need

There is a lot we may be missing as altmetrics researchers and services if we primarily concern ourselves with counting citations, grant dollars, patents, altmetrics, and other numeric indicators to understand the value of research.

I would like to see the field do more to understand: (a) the human motivations that *create* altmetrics (e.g., Why do people peer review some works but not others on Pubpeer?), (b) how the nuances of language and the limits of current technology affect the altmetrics studies we are able to do, and (c) the implicit and explicit values that underpin the types of data we collect and use in evaluation. Put as a series of research questions, we might ask:

- Why do people share and discuss research on various social media sites?
- How should we account for the nuances of language and limits of technology in how we approach certain kinds of altmetrics studies?
- What values are embedded in the institutions using altmetrics?

We already know why some people use Mendeley to bookmark research [11], but this survey-backed study is the exception rather than the rule in altmetrics research. More research is needed that directly asks those sharing scholarship on other online platforms their motivations for doing so. That will help us better estimate what, for example, a metric like "19 open peer reviews about an article" might mean if we know that 75% of peer reviewers tend to mention research in order to criticize it.

Survey-based approaches to understanding motivation are needed because manual coding of the mentions that research receives online (as seen in [12–15]) simply cannot scale. With more than 60 million mentions of 9 million research outputs collected since 2011, human coders cannot keep up with the pace of discussion across scores of research disciplines, languages, and demographics.

One might be inclined to suggest that computers could be used to analyze altmetrics mentions instead, giving us a scalable way to interpret the meanings of tweets, news mentions, policy citations, and other online discussions. But sentiment analysis technology is not yet able to make crucial, reliable distinctions in interpreting the intricacies of language like irony or sarcasm (especially for non-Western European languages), making this approach unviable for the time being.

Our last and perhaps greatest challenge is a cultural one, rather than a strictly technological one. For all the attention paid to institutional values in grand university strategic plans and workplace "culture" within the Academy, many evaluation practices at universities worldwide are surprisingly out of step with the question of values, period. Or, institutional values don't align with the values of rank-and-file faculty and students, causing frustration and mistrust.

Take for example the practice of measuring researchers' "productivity" by counting the number of papers they publish early on in their careers. This practice is founded upon the implicit values of competition and industriousness - the idea that the constant production of tangible new content like journal articles in order to prove how much harder you have been working than others is something to be valued over and above the often-invisible work of deeply engaging with a topic. Researchers argue that counting publications has led to the practice of "salami slicing", and that a focus on writing takes away from the crucial work of doing the research itself [16, 17]. Though clearly harmful to the culture of research, this practice persists.

Other positive values can and should drive the academic endeavor, as they will lead to evaluation practices that can incentivize positive behaviors within the academy[1]. For example, the value of "openness" defined as making one's research as accessible as possible to others both within and beyond academia, might take the form of only allowing Open Access publications to be considered in evaluation scenarios such as the UK Research Excellence Framework. Such Open Access publishing and self-deposit mandates have been shown to increase the percentage of research that is publicly available. Clearly, incentivization aligned with well-articulated values, though not without its limitations [18] can produce desirable outcomes that stand to benefit all researchers.

One could also take a closer look at the value of "quality". This is potentially one of the only values that's universal within academia; everyone wants to produce (or wants their faculty to produce) high-quality scholarship.

The traditional means of understanding "quality" are quite limited: Journal Impact Factors are seen to be unreliable for judging article-level impact [19]; counting raw citations can be a crap-shoot given the many meanings of citations [20, 21]; and subjective evaluation practices like peer review are (a) unscalable, and (b) arguably subject to bias [22].

One could easily think of new kinds of qualitative and quantitative evidence to understand the quality of research, if one breaks down the components of "quality" that are often important to evaluators: reproducibility, creativity, intentionality, and the

---

[1] www.humetricshss.org.

advancement of knowledge. A few examples of traditional and nascent evidence include:

- Quality: book reviews, peer reviews
- Reproducibility: cited in Methods, data analysis code "forked" on Github
- Creativity: interdisciplinary citing, new formats of research, depth of elaboration
- Advancing knowledge: sustained citations over time, awards, sustained social media discussions
- Intentionality: time spent/depth of thinking, regular reflection upon goals

These kinds of evidence could easily be listed in tenure and promotion dossier preparation instructions, as well as instructions for reviewers, to signal to researchers that there are many paths to understanding "quality". Imagine the burden that would lift from the shoulders of faculty who work in obscure, rarely-cited, low Journal Impact Factor fields. All of that said, much of the research and app development in altmetrics avoids the question of values altogether. The altmetrics community should, in my opinion, get more comfortable with the idea of developing indicators that signal one's progress towards embodying certain values. Providing better, more accurate indicators that are grounded in values is likely the best path towards changing an institutional culture that is, for better or for worse, reliant upon metrics for decision-making.

4 What the Altmetrics Community Can Do for Greater Context

In conclusion, there are a number of context-producing actions that the altmetrics community can take to improve evaluation in academia.

Apps makers like Altmetric and Impactstory can provide more context in the data and reports they display. For example, accurate subject area benchmarks can help evaluators understand if the number of policy citations a paper receives is above average (or the opposite). Reporting of raw counts will no longer do.

Apps makers can also develop tools that better account for language. Sentiment analysis of the tweets that a paper receives could signal potential issues with article quality, for example. Adding "sentiment scores" to other altmetrics could help evaluators quickly identify papers that warrant closer attention. Though sentiment analysis technologies are not yet perfected, with advances in machine learning, it's not impossible to expect that we can get a lot better at detecting irony and other types of sentiment analysis work and presenting it for interpretation in easily understood ways.

Apps makers should also be working towards moving from "data" to "wisdom". Currently, most altmetrics apps provide a great deal of raw data to evaluators without necessarily offering insight into the value of a body of work. We can and should do more to highlight the data that can tell compelling stories about the impact of one's research, in an elegant, usable, and easily understandable way. We're working on all of these problems at Altmetric, and we believe other vendors in the field should be, too.

Researchers like you have a similarly important role to play in the betterment of the field. I encourage the community to continue to study the motivations behind the actions that lead to altmetrics. Help us answer definitively: Why is research cited in public policy? How do people talk about research on social media? Knowing the answers to questions like these can in turn help app developers create more precise indicators and features that offer insight on the impact of research.

Researchers should also consider how their research will be applied in the real world. Knowing that evaluators often favor metrics over impact qualitative evidence, can the community develop better indicators that are less prone to abuse? Can we discover more exact and useful "baskets of metrics" [23] that more precisely pinpoint various "flavours of impact" [24]? What can we learn from evaluators themselves that might improve the research we do [25]? I do believe it is possible.

Finally, there is the related question of accounting for context in evaluation, including institutional values, the prevalence of an environment of scarcity within the academy, and how metrics can incentivize certain behaviors in practice. What role can and should the altmetrics research community play in encouraging open discussion of values within academia, so that we might develop better indicators to help researchers and institutions alike measure their progress towards embodying those values?

Let's do more as a community that cares deeply about altmetrics to encourage greater context in research evaluation.

References

1. Adie, E.: Don't throw out the baby with the metrics bathwater (2015). https://www.timeshighereducation.com/blog/dont-throw-out-baby-metrics-bathwater
2. Alperin, J.P.: Geographic variation in social media metrics: an analysis of Latin American journal articles. Aslib J. Inf. Manag. **67**, 289–304 (2015)
3. Konkiel, S.: How research is shared on VKontakte (2016). https://www.altmetric.com/blog/vkontakte/
4. Wang, X., Fang, Z., Li, Q., Guo, X.: The poor altmetric performance of publications authored by researchers in mainland china. Front. Res. Metrics Anal. **1**, 8 (2016)
5. Zahedi, Z.: What explains the imbalance use of social media across different countries? A cross country analysis of presence of Twitter users tweeting scholarly publications. In: 4th Altmetrics Conference, Toronto, Canada (2017)
6. Sugimoto, C.R., Larivière, V.: Altmetrics: broadening impact or amplifying voices? ACS Cent. Sci. **3**, 674–676 (2017)
7. Science, D., Charman-Anderson, S., Kane, L., Meadows, A., Greshake Tzovaras, B., Konkiel, S., Wheeler, L.: Championing the Success of Women in Science, Technology, Engineering, Maths, and Medicine (2017)
8. Alhoori, H., Ray Choudhury, S., Kanan, T., Fox, E., Furuta, R., Giles, C.L.: On the relationship between open access and altmetrics. In: iConference 2015 Proceedings (2015)
9. Erdt, M., Aung, H.H., Aw, A.S., Rapple, C., Theng, Y.-L.: Analysing researchers' outreach efforts and the association with publication metrics: A case study of Kudos. PLoS ONE **12**, e0183217 (2017)
10. Adie, E.: Gaming altmetrics (2013). https://www.altmetric.com/blog/gaming-altmetrics/
11. Mohammadi, E., Thelwall, M., Kousha, K.: Can mendeley bookmarks reflect readership? A survey of user motivations. J. Assoc. Inf. Sci. Technol. **67**, 1198–1209 (2016)
12. Bowman, T.D.: Differences in personal and professional tweets of scholars. Aslib J. Inf. Manag. **67**, 356–371 (2015)

13. Friedrich, N., Bowman, T.D., Stock, W.G., Haustein, S.: Adapting sentiment analysis for tweets linking to scientific papers (2015)
14. Tsou, A., Bowman, T., Ghazinejad, A., Sugimoto, C.: Who Tweets about Science ? In: Proceedings of the ISSI (2015)
15. Shema, H., Bar-Ilan, J., Thelwall, M.: How is research blogged? A content analysis approach. J. Assoc. Inf. Sci. Technol. **66**, 1136–1149 (2015)
16. Abbott, A., Cyranoski, D., Jones, N., Maher, B., Schiermeier, Q., Van Noorden, R.: Metrics: do metrics matter? Nature **465**, 860–862 (2010)
17. Haustein, S., Larivière, V.: The use of bibliometrics for assessing research: possibilities, limitations and adverse effects. In: Welpe, I.M., Wollersheim, J., Ringelhan, S., Osterloh, M. (eds.) Incentives and Performance, pp. 121–139. Springer, Cham (2015). https://doi.org/10.1007/978-3-319-09785-5_8
18. Hammarfelt, B.M.S., de Rijcke, S., Rushforth, A.D.: Quantified academic selves: the gamification of science through social networking services. Inf. Res. **21** (2016)
19. Smith, R.: Commentary: the power of the unrelenting impact factor—Is it a force for good or harm? Int. J. Epidemiol. **35**, 1129–1130 (2006)
20. Cronin, B.: The Citation Process : The Role and Significance of Citations in Scientific Communication. T. Graham (1984)
21. Leydesdorff, L., Bornmann, L., Comins, J.A., Milojević, S.: Citations: indicators of quality? the impact fallacy. Front. Res. Metrics Anal. **1**, 1 (2016)
22. Lee, C.J., Sugimoto, C.R., Zhang, G., Cronin, B.: Bias in peer review. J. Am. Soc. Inf. Sci. Technol. **64**, 2–17 (2013)
23. Wilsdon, J., Allen, L., Belfiore, E., Campbell, P., Curry, S., Hill, S., Jones, R., Kain, R., Kerridge, S., Thelwall, M., Tinkler, J., Viney, I., Wouters, P., Hill, J., Johnson, B.: The Metric Tide: Report of the Independent Review of the Role of Metrics in Research Assessment and Management (2015)
24. Piwowar, H.: Altmetrics shows that citations can't stand up to the full 31 flavours of research impact. Impact of Social Sciences (2012). http://blogs.lse.ac.uk/impactofsocialsciences/2012/04/04/31-flavours-research-impact/
25. de Rijcke, S., Rushforth, A.: To intervene or not to intervene; is that the question? On the role of scientometrics in research evaluation. J. Assoc. Inf. Sci. Technol. **66**, 1954–1958 (2015)

Full Papers

Monitoring the Broader Impact of the Journal Publication Output on Country Level: A Case Study for Austria

Juan Gorraiz(✉) ⓘ, Benedikt Blahousⓘ, and Martin Wielandⓘ

Library and Archive Services, Department for Bibliometrics and Publication
Strategies, University of Vienna, Boltzmanngasse 5, 1090 Vienna, Austria
juan.gorraiz@univie.ac.at

Abstract. This study provides an example of a monitoring practice concerning
the broader impact of the journal publication output on country level. All Austrian
publications of the last three publication years indexed in WoS Core Collection
and mapped to a DOI were analysed in PlumX. The metrics traced in the different
data sources were compared for six main knowledge areas. The results reinforce
the importance of the usage metrics especially in disciplines related to the area
"Arts & Humanities". The highest data coverage is provided by the number of
readers in Mendeley. The percentage of publications with social media scores,
especially tweets, has been significantly increasing within the last three years, in
agreement with the increasing popularity of these tools in recent years. The highest
values for social media are reported for the Health and Life Sciences, followed
very closely by the Social Sciences. The relative insignificance in the Arts &
Humanities' is noteworthy. Our study confirms very low correlation values
between the different measures traced in PlumX and supports the hypothesis that
these metrics should rather be considered as complementary sources. High correlations between the same measures or metrics originating from different data
sources were only reported for citations, but not for usage data. Medium correlation values were observed between usage and citation counts in WoS Core
Collection. No association of the number of co-authors or co-affiliations with any
of the measures considered in this study could be found, except for a low correlation between the number of affiliations and captures or citations.

Keywords: New metrics · Altmetrics · Citation analysis · PlumX
Web of Science · Impact · Monitoring

1 Introduction

Two highly revolutionary developments have impacted the early 21st century [1]. The
first was the rapid adoption of digitally available information on the web, particularly
e-journals [2, 3]. Due to the increasing availability, the tracking and collection of usage
data (e.g. views and downloads) has become much easier than in the print-only era.
This consequently resulted in a renaissance of the usage metrics, which have become
increasingly popular beyond librarian practices and can now be used in scientometric
analyses as complementary data to citation metrics [4]. The second revolution was the

© Springer Nature Singapore Pte Ltd. 2018
M. Erdt et al. (Eds.): AROSIM 2018, CCIS 856, pp. 39–62, 2018.
https://doi.org/10.1007/978-981-13-1053-9_4

evolution of the internet into a social media platform and the vast adoption of Web 2.0 practices, even if still strongly influenced by demographic characteristics, such as age and gender, but also by position and discipline [5–8].

New social media-based metrics have now evolved along with traditional and non-traditional metrics. This broader range of metrics now allows a more diverse assessment of research outputs on different levels. However, the use of such new metrics for evaluative purposes is still very controversial and is certainly attached to many challenges. Reliability, completeness, interrelationship, standardization, stability, scalability and normalization of the collected data are still unresolved issues to be tackled. Certainly the use of new metrics should not only be reduced to evaluative purposes. They are certainly a popular and very easy-to-use means for scientists, institutions or publishers alike to promote their research output and to enhance their visibility. Thereby they also increase the potential for higher impact [1].

Thus, all academic sectors and players involved in scientific communication need to rise to this new challenge and confront it somehow. Especially, modern scientific libraries are obliged to face the new challenges of the digital era rather sooner than later in a professional way. The use of new metrics is an emerging field for academic libraries. Responsible use and qualified practices will offer many opportunities to provide innovative services specifically tailored to academic and administrative communities in order to: (1) face the new challenge and to stimulate a positive attitude towards the use of the new metrics; (2) support scientists in the 'publish or perish' dilemma: planning a scientific career and developing adequate publication strategies (especially for young scientists); (3) enhance the visibility: the institutional one (Rankings & Web presence) as well as the individual one (adoption of permanent identifiers, help and assistance in the promotion game) and (4) prevent administrations from bad use of these new metrics and incorrect interpretations (informed peer review) [9].

New metrics are indispensable in the daily life of modern libraries challenged by continuously changing demands. This does not only include the support of strategic decisions in licensing and collection management. It is likewise important for the enrichment of research documentation systems and repositories, the monitoring of policies (e.g., Open Access policies), and the development of new indicators. The latter is particularly true for disciplines where traditional bibliometrics fail to work, like in the humanities. The monitoring of adopted policies (e.g., Affiliation policy, Open Access policy, etc.) and of the institutional web impact belongs to the services, which our department currently offers[1].

The aim of this paper is to provide an example of a monitoring practice concerning the broader impact [10] of the journal publication output on country level and to discuss the opportunities and limitations when providing similar reports. Therefore, we analyze the publications of three years by means of traditional and new metrics, including altmetrics, and discuss their possible interpretation. Furthermore, we analyze which disciplines (or knowledge areas) are the most present and active according to the different metrics and their evolution within the three last more recent years[2]. Finally, we explore if there is some correlation, first, between the metrics (usage and citation)

---

[1] See http://bibliothek.univie.ac.at/bibliometrie/services.html.

[2] Data were gathered by the end of December 2017.

collected in different data sources (WoS Core Collection and PlumX) and second, between the intensity of the signals or scores retrieved for different dimensions or categories (mentions, captures, citations and usage data) retrieved in the same data source (in this case via PlumX). Last but not least, the association of the number of co-authors or co-affiliations with the intensity of all the measures and metrics traced in this study has been analyzed.

2 Data Samples and Methodology

We use the Web of Science Core Collection including all the comprised indices (proceedings, books, etc.) as our data source. All publications containing "Austria" at least once in their affiliation data were retrieved and downloaded for the three complete publication years: 2014, 2015 and 2016. For this purpose, we used the search string "CU = (Austria)" in the advanced search, and then selected and downloaded all the publications with available DOIs. The number of documents indexed in WoS Core Collection with DOI rose from 71.8% in 2014 to 78.7% in 2016. Most of the documents without DOI were meeting abstracts (90% without DOI). Only considering articles as document type, the percentage of documents with DOI increased from 93.5% in 2014 to 94.4% in 2016. These results are in agreement with prior studies (e.g., [11]).

For all metrics data collection and aggregation, we have used the fee-based PlumX altmetrics dashboard because it gathers and offers publication-level metrics for so-called artifacts, which also include different document types, and it allows DOIs to be directly used as well as many other identifiers (IDs). The provider of PlumX is Plum Analytics, a 100% subsidiary of EBSCO Information Services since 2014. However, by the end of 2016, Elsevier took over PlumX from EBSCO.

In order to gather the data in PlumX, a plain text file containing all the DOIs for all publications retrieved in WoS Core Collection has been uploaded to PlumX and processed by the tool, providing a new dataset including all the resulting "artifacts" - as data records are named in PlumX - and the corresponding altmetric scores gathered from each tool covered by PlumX. The resulting dataset for each data record type has been exported to Excel in CSV format, and then analyzed for each year (2014, 2015 and 2016). The processed datasets for each year are described in Table 1. It should be noted that more than 90% of the records were correctly identified in PlumX.

Table 1. Processed datasets in PlumX

Input		Output		
Dataset	# records	Dataset	# records	% records availability
WoS CC 2014	16626	PlumX 2014	16538	99.47%
WoS CC 2015	18615	PlumX 2015	18486	99.31%
WoS CC 2016	20631	PlumX 2016	20452	99.13%

The resulting dataset also includes the scores of all measures according to their origin. The measures are categorized into five separate dimensions: usage, captures, mentions, social media, and citations [12]. This categorization may be subject to criticism, but one big advantage of PlumX is that the results are differentiated in the resulting dataset for each measure, and its origin and can be aggregated according to the user criterion.

In order to analyze the differences between knowledge areas, all publications retrieved in WoS Core Collection have been reclassified according to the field "research areas" in six main knowledge areas: (1) Life Sciences, (2) Physical Sciences, (3) Engineering & Technology, (4) Health Sciences, (5) Social Sciences, and (6) Arts & Humanities.

2.1 Metrics Comparisons in WoS CC and PlumX

For all Austrian publications with DOI indexed in WoS CC, statistical and correlation analyses were performed for the number of citations indexed in Web of Science Core Collection (field TC in WoS data export) and in the whole WoS platform (field ZA in WoS data export) versus the number of citations attracted in Scopus and CrossRef according to PlumX. For the other citation indexes collected via PlumX, the data sample was not large enough to perform a sound correlation analysis.

Furthermore, the results of the usage metrics, included since 2015 in WoS Core Collection (via Clarivate Analytics), have been correlated with the ones provided by PlumX (via EBSCO). According to Clarivate, the WoS usage metric reflects the number of times the article has met a user's information needs as demonstrated by clicking links to the full-length article at the publisher's website (via direct link or Open-URL) or by saving the article for personal use in a bibliographic management tool (via direct e port or in a format to be imported later[3]). In our study, both indicators, U1 (= the count of the number of times the full text of a record has been accessed or a record has been saved within the last 180 days) and U2 (= the count of the number of times the full text of a record has been accessed or a record has been saved since February 1, 2013) have been compared with the corresponding number of abstract views traced by PlumX via EBSCO.

2.2 Correlations Between Different Metrics in Each Data Source

Spearman correlations for all PlumX measures with a significant number of data have been performed for each publication year. These are - number of readers in Mendeley (captures), (number of citations in Scopus (citations), numbers of tweets in Twitter (social media), number of Abstract and HTML/PDF views in EBSCO (usage).

A correlation analysis was also performed in WoS Core Collection for citation and usage counts traced exclusively in this tool.

Finally, correlations were computed between the number of authors or number of affiliations and the most representative measures traced in each data source: citation and

---

[3] See also https://images.webofknowledge.com/images/help/WOS/hp_usage_score.html.

usage counts in WoS Core Collection, number of readers in Mendeley, citations in Scopus, number of tweets in Twitter and abstracts views in EBSCO in PlumX. The results were then compared to the ones obtained in previous studies.

3 Results

The results of the analyses are presented in the following sections.

3.1 General Results

The complete results from PlumX for the years 2014, 2015 and 2016 are summarized in the Appendix as Tables 11, 12 and 13 respectively and they include the following information:

- data availability = number of data records traced in PlumX: see information at the top of each table
- data with scores = number of data records traced in PlumX with at least one score (>=1)
- data with scores (%) = number of data records traced in PlumX with at least one score (>=1) in relation to the number of the WoS CC records searched
- intensity = sum of all signals or scores
- mean = numerical mean of all samples
- density (mean available) = sum of all signals or scores in relation to the number of all WoS CC records traced in PlumX with at least one score (>=1)
- median = median of all samples
- maximum = maximal number of signals or scores
- standard deviation
- T confidence interval = with a value of 0.05 for the variable α;

"Data availability" reflects that data records could be found in PlumX although they may not have any altmetric signals or scores to analyze (i.e., similar to uncitedness). "Data with scores" shows the number of data records that have received at least one score or signal. We refer to the percentage of documents with at least one score as coverage or degree of availability (e.g., via tweeting a DOI).

As explained by Haustein et al. [13], intensity describes how often a data record has been referenced on a social media platform. Density resembles the citation rate that is highly affected by the "data with scores" (e.g., low data with scores leads to poor values for density).

Furthermore, the measures (signals or scores) are categorized in five separate dimensions typical for PlumX data: usage, captures, mentions, social media, and citations. It should be noted that the total values for each dimension were only calculated in order to give a quick overview of the percentage of documents with available data. Nevertheless, the dimensions reflect different types of engagement with collected data records that should not be conflated. Tables 11 and 13 (refer Appendix) show the large diversity of measures traced for each measure in each dimension. They also show that the distribution of the collected signals or scores for each measure is obviously highly skewed as evident from the statistical analysis also included in these tables.

It is noteworthy that only measures or tools, where data could be traced, are included in these tables. For example, measures like "catalogue holdings" or "reviews", reported in previous studies [12] as most characteristic for books, are missing in this study. This is explained by the structure of our dataset (more than 90% of the publications with DOI retrieved in WoS CC were journal articles and reviews) as well as by the use of the DOIs, and not ISBNs, in order to trace the corresponding scores in PlumX. Total score calculations for any dimension or categorical group were only used as an aid in order to quickly estimate the data availability and coverage percentage for each dimension. However, due to the heterogeneous mix of data and aspects, mathematical sums were not used for calculating correlations.

Furthermore, the study also describes the changes in the values of the most significant measures and indicators over time. Table 2 shows a summary of the most relevant facts (number of publications, coverage, intensity and density) for each dimension (captures, citations, mentions, social media and usage) and for the more representative measures for each one (e.g. number of Mendeley readers as captures, tweets as social media, etc.). It is expected that the percentage of coverage or data availability, the intensity and density decreases for each metric or measure according to the publication year of the documents, due to the decreasing measured time windows (three, two and one year for publications of the years 2014, 2015 and 2016 respectively). This is clearly observed for three dimensions (captures, citations and usage), but it is just the opposite for mentions and for social media.

The results show that the percentage of Austrian WoS CC publications with mentions increases rather slowly (2%), while the percentage of publications with social media scores has been steadily increasing from $\sim$28% in 2014 to $\sim$38% in 2016. Even the tweets intensity has been increasing within the last years despite the reduced time window. Concerning intensity and density, the results show that usage counts are quantitatively predominant in comparison to the other metrics. Abstract views are responsible for the highest values followed by captures (number of readers in Mendeley).

3.2   Results According to the Six Main Research Areas

Table 3 informs about the coverage (percentage of data with scores) in total and for each dimension for the publications from each year. The lowest percentage of total coverage is reported by the Arts & Humanities area. Nevertheless, usage scores are still high. Actually, the percentage of usage data is higher than in the other four dimensions. The percentage of uncited data in this area is however twice or three times higher than in the other four considered areas. This is due to the longer citing half-life and the lower reference densities characteristic for this discipline. The behavior of the four hard sciences (in this case, Engineering & Technology, Health Sciences, Life Sciences and Physical Sciences) is very similar in all dimensions, except in the social media dimension, where Health and Life Sciences account for the highest percentages of data availability, followed very closely by the Social Sciences. The percentage values for Social Sciences in the other three dimensions (captures, citations and mentions) are in-between the ones for the hard sciences and the ones for Arts & Humanities. A low

Table 2. Summary of the results for the three publication years and most active measures

	2014				2015				2016			
	Items with data scores >= 1	# records with data scores >=1/# WoS CC Records (%)	Intensity (Sum)	Density (Mean available)	Items with data scores >= 1	# records with data scores >=1/# WoS CC Records (%)	Intensity (Sum)	Density (Mean available)	Items with data scores >= 1	# records with data scores >=1/# WoS CC Records (%)	Intensity (Sum)	Density (Mean available)
Readers: Mendeley	15299	92.5	342723	22.4	16780	90.8	301235	18.0	17857	87.3	240296	13.5
Total captures	15622	94.5	464318	29.7	17017	92.1	392835	23.1	18116	88.6	296880	16.4
Citations: Scopus	14058	85.0	201533	14.3	14544	78.7	133910	9.2	12863	62.9	64197	5.0
Citations: CrossRef	13515	81.7	175586	13.0	14348	77.6	124838	8.7	12826	62.7	61783	4.8
Total citations	14549	88.0	418165	28.7	15367	83.1	285940	18.6	14175	69.3	137644	9.7
Tweets: Twitter	4198	25.4	38161	9.1	5726	31.0	64695	11.3	7503	36.7	73031	9.7
Shares Likes & Comments: Facebook	1485	9.0	83666	56.3	1572	8.5	92378	58.8	1525	7.5	79026	51.8
Total social media	4715	28.5	123369	26.2	6128	33.1	158231	25.8	7824	38.3	152385	19.5
Blog mentions	339	2.0	1567	4.6	456	2.5	964	2.1	500	2.4	1147	2.3

(continued)

Table 2. (*continued*)

	2014				2015				2016			
	Items with data scores >= 1	# records with data scores >=1/# WoS CC Records (%)	Intensity (Sum)	Density (Mean available)	Items with data scores >= 1	# records with data scores >=1/# WoS CC Records (%)	Intensity (Sum)	Density (Mean available)	Items with data scores >= 1	# records with data scores >=1/# WoS CC Records (%)	Intensity (Sum)	Density (Mean available)
Links: Wikipedia	359	2.2	517	1.4	344	1.9	756	2.2	250	1.2	383	1.5
News mentions	225	1.4	1166	5.2	615	3.3	3494	5.7	946	4.6	3940	4.2
Total mentions	815	4.9	6999	8.6	1183	6.4	7208	6.1	1425	7.0	6313	4.4
EBSCO: Abstract views	14650	88.6	2048497	139.8	15318	82.9	1700018	111.0	16271	79.6	1369298	84.2
PDF views	838	5.1	43925	52.4	1	0.0	56	56.0	3	0.0	3156	1052.0
HTML views	4463	27.0	284700	63.8	414	2.2	799679	1931.6	369	1.8	604071	1637.0
Total usage	14721	89.0	5290701	359.4	15430	83.5	3373836	218.7	16449	80.4	2546079	154.8
Total all	16203	98.0	6303552	389.0	17981	97.3	4218050	234.6	19596	95.8	3139301	160.2

Table 3. Coverage in PlumX for each measure, publication year and main knowledge area.

PY	Subject category	Total number of items	% of items with scores	% of items with scores in captures	% of items with scores in citations	% of items with scores in social media	% of items with scores in mentions	% of items with scores in usage
2014	Arts & Humanities	143	91.6%	65.0%	35.7%	7.0%	2.1%	91.6%
	Engineering & Technology	3163	97.7%	94.0%	88.5%	20.9%	5.2%	80.7%
	Health Sciences	4896	97.7%	95.0%	86.4%	37.1%	5.8%	94.1%
	Life Sciences	3493	98.9%	96.5%	92.5%	36.4%	6.1%	92.0%
	Physical Sciences	3560	99.0%	94.2%	90.5%	14.7%	2.4%	86.0%
	Social Sciences	1260	95.2%	92.5%	79.2%	33.7%	4.9%	90.6%
	Not available	1	100.0%	100.0%	100.0%	0.0%	0.0%	0.0%
	Total	16516	98.0%	94.5%	88.0%	28.5%	4.9%	89.0%
2015	Arts & Humanities	168	85.1%	61.9%	26.8%	7.1%	1.8%	79.2%
	Engineering & Technology	3964	96.0%	90.3%	81.0%	23.8%	6.3%	69.2%
	Health Sciences	5288	97.3%	92.9%	82.5%	44.5%	6.9%	92.0%
	Life Sciences	3616	98.9%	96.2%	90.7%	45.9%	10.0%	91.8%
	Physical Sciences	3820	97.9%	90.3%	85.9%	16.5%	3.1%	76.9%
	Social Sciences	1574	96.1%	91.4%	72.4%	33.2%	5.5%	88.1%
	Not available	22	100.0%	95.5%	81.8%	0.0%	0.0%	86.4%
	Total	18452	97.3%	92.1%	83.1%	33.1%	6.4%	83.5%
2016	Arts & Humanities	218	85.8%	46.8%	16.1%	9.2%	2.8%	81.2%
	Engineering & Technology	4529	94.7%	89.2%	65.2%	27.8%	6.8%	70.2%
	Health Sciences	5724	96.1%	88.6%	68.8%	50.5%	7.6%	88.0%

(continued)

Table 3. (*continued*)

PY	Subject category	Total number of items	% of items with scores	% of items with scores in captures	% of items with scores in citations	% of items with scores in social media	% of items with scores in mentions	% of items with scores in usage
	Life Sciences	4103	98.3%	93.9%	78.2%	54.1%	10.6%	88.2%
	Physical Sciences	3991	96.2%	86.5%	74.8%	19.9%	3.9%	72.2%
	Social Sciences	1696	92.5%	85.5%	57.4%	35.3%	4.5%	86.3%
	Not available	122	91.0%	75.4%	24.6%	15.6%	0.8%	47.5%
	Total	20383	95.8%	88.6%	69.3%	38.3%	6.9%	80.5%

percentage of data availability was observed for the Physical Sciences in the social media dimension. The validity of these results is corroborated by the data resulting for each publication year. In general, all percentages are decreasing according to the decreasing measuring window, except for the social media dimension, where values are significantly increasing in all six knowledge areas.

Table 4 informs about the absolute number of items with scores and the density (sum of all signals or scores in relation to the number of all WoS records traced in PlumX with at least one score ($>=1$)) for the publications from each year and for each area. In this case, we used the most representative measures of each dimension instead of the total sum of signals due to the heterogeneity of the data collected in each tool (as already mentioned above): number of readers in Mendeley (captures), citations from Scopus, number of tweets in Twitter (social media) and number of abstract views, PDF Views and HTML Views in EBSCO. The number of data collected (intensity) as well as the density resulting from the mentions was not significant enough to be considered in this further analysis (see Supplementary Material).

The density results correspond very well to the ones concerning data coverage or availability. High usage density values in the Arts & Humanities and Social Sciences and lower ones for Physical Sciences could be observed in comparison to the other hard sciences. In this case all densities are decreasing according to the reduced measuring window.

3.3   Results from the Citation Analysis

Table 5 shows the number of records lost through data matching when checking the WoS CC data introduced in PlumX and the resulting dataset. It illustrates the difficulty and maximal accuracy matching the data of the data input and output in PlumX.

A summary of the descriptive statistics for citation counts in the four consulted data sources (WoS Core Collection, WoS (overall platform), Scopus and CrossRef) is provided in Table 6. They show very similar values in all four data sources. The citation

Table 4. Density of the most representative measures traced in PlumX for each publication year and main knowledge area.

PY	Subject area	Number of items with captures: Readers: Mendeley	Density: Captures: Readers: Mendeley	Number of items with citations: Scopus	Density: Citations: Scopus	Number of items with social media: Tweets: Twitter	Density: Social media: Tweets: Twitter	Number of items with usage: Abstract views: EBSCO	Density: Usage: Abstract views: EBSCO
2014	Arts & Humanities	68	8.47	46	2.63	10	2.20	130	264.16
	Eng & Technology	2936	25.36	2706	13.32	563	21.06	2540	166.16
	Health Sciences	4554	20.99	4103	17.82	1588	7.34	4590	101.85
	Life Sciences	3346	29.65	3137	13.82	1182	8.72	3197	106.49
	Physical Sciences	3261	13.07	3115	12.76	471	2.66	3038	61.21
	Social Sciences	1116	26.98	937	9.61	377	8.05	1140	524.16
2015	Arts & Humanities	71	5.61	43	2.44	12	2.92	133	195.46
	Eng & Technology	3546	19.20	3038	8.24	845	32.69	2715	126.63
	Health Sciences	4854	17.34	4168	11.61	2174	9.34	4848	84.52
	Life Sciences	3450	23.98	3137	8.96	1599	6.34	3299	105.19
	Physical Sciences	3423	11.50	3095	8.46	584	5.07	2904	44.88
	Social Sciences	1385	18.57	1017	5.57	500	7.10	1377	320.83

(continued)

Table 4. (*continued*)

PY	Subject area	Number of items with captures: Readers: Mendeley	Density: Captures: Readers: Mendeley	Number of items with citations: Scopus	Density: Citations: Scopus	Number of items with social media: Tweets: Twitter	Density: Social media: Tweets: Twitter	Number of items with usage: Abstract views: EBSCO	Density: Usage: Abstract views: EBSCO
2016	Arts & Humanities	66	3.09	32	1.31	16	8.13	177	101.19
	Eng & Technology	4009	14.61	2675	4.49	1176	15.34	3147	100.91
	Health Sciences	4978	12.66	3558	5.77	2782	9.39	4985	57.84
	Life Sciences	3820	17.31	2975	4.78	2168	9.34	3593	79.23
	Physical Sciences	3434	9.11	2766	5.30	746	3.58	2827	38.27
	Social Sciences	1400	14.10	796	2.85	574	7.68	1454	234.86

Table 5. Number of records after data matching in WoS CC and PlumX.

Year	WoS-CC records in PlumX	Records in PlumX output	Records with matching record in PlumX (via DOI)	Share of records with matching record in PlumX (via DOI)	Records without matching record in PlumX (via DOI)	Without matching record in WoS-CC (via DOI)
2014	16,664	16,538	16,516	99.11%	148	22
2015	18,615	18,486	18,452	99.12%	163	34
2016	20,639	20,452	20,383	98.76%	256	69

intensity (total number of citations) in Scopus is higher than in Web of Science as expected and due to the higher number of journals indexed. The lower citation intensity in CrossRef is not that obvious, given that all the publications considered in this study have Digital Object Identifiers (DOIs). It has nonetheless also been reported in recent studies performed for journal articles uploaded to Zenodo [14].

Table 6. Descriptive statistics of the citation analysis performed in WoS CC and PlumX

PY	Parameter	WoS-CC TC	WoS-CC Z9	PlumX citations: Scopus	PlumX citations: CrossRef
2014	Total number of items	16,516	16,516	16,516	16,516
	Number of uncited items	2,526	2,453	2,472	3,017
	Share of uncited items	15%	15%	15%	18%
	Total number of citations	175,332	180,431	201,395	175,465
	Mean number of citations	10.62	10.92	12.19	10.62
	Mean (cited items only)	12.53	12.83	14.34	13
	Median of citations	5	5	6	5
	Maximum of citations	1,418	1,487	2,563	1,457
	Standard Deviation	26.84	27.60	37.11	28.09
	T- Confidenz (alpha=0.05)	0.41	0.42	0.57	0.43

(*continued*)

Table 6. (*continued*)

PY	Parameter	WoS-CC TC	WoS-CC Z9	PlumX citations: Scopus	PlumX citations: CrossRef
2015	Total number of items	18,452	18,452	18,452	18,452
	Number of uncited items	4,140	4,067	3,936	4,135
	Share of uncited items	22%	22%	21%	22%
	Total number of citations	115,717	118,629	133,565	124,535
	Mean number of citations	6.27	6.43	7.24	6.75
	Mean (cited items only)	8.09	8.25	9.20	8.70
	Median of citations	3	3	3	3
	Maximum of citations	1,243	1,302	1,509	1,332
	Standard Deviation	18.51	19.05	21.88	19.93
	T- Confidenz (alpha=0.05)	0.27	0.27	0.32	0.29
2016	Total number of items	20,383	20,383	20,383	20,383
	Number of uncited items	8,604	8,516	7,558	7,597
	Share of uncited items	42%	42%	37%	37%
	Total number of citations	51,748	52,780	63,783	61,560
	Mean number of citations	2.54	2.59	3.13	3.02
	Mean (cited items only)	4.39	4.45	4.97	4.81
	Median of citations	1	1	1	1
	Maximum of citations	564	584	527	505
	Standard Deviation	8.43	8.59	9.25	9.19
	T- Confidenz (alpha=0.05)	0.12	0.12	0.13	0.13

Table 7. Citation correlations WoS CC and PlumX (via Scopus and CrossRef)

Year	WoS CC: Times cited (TC and Z9)	PlumX citations: Scopus	PlumX citations: CrossRef
2014	TC	0.93	0.93
2015	TC	0.98	0.97
2016	TC	0.95	0.96
2014	Z9	0.93	0.93
2015	Z9	0.98	0.97
2016	Z9	0.95	0.95

The results of the Spearman correlations performed for the four citation counts are summarized in Table 7. They show a very high correlation between the citations counted in the three data sources. As expected, the correlations between citation counts in WoS and in Scopus via PlumX are very high and in agreement with previous results [15, 16]. However, the correlation with CrossRef is much higher as reported in the last mentioned study [14]. This can be due to the short citation window and larger sample sizes in our study, but needs further exploration.

Table 8 shows the correlations obtained for the usage data provided in WoS Core Collection and PlumX via EBSCO (abstracts views). The reported correlations are low in comparison to the ones calculated for the citation measures or indicators. This could hint at the fact, that the user groups are different in both data sources and have different interests.

Table 8. Usage correlations WoS U1 and U2 versus PlumX-Abstracts views via EBSCO

Year	WoS CC usage metrics	PlumX: Abstract views: EBSCO
2014	U1	0.13
2015	U1	0.21
2016	U1	0.19
2014	U2	0.17
2015	U2	0.20
2016	U2	0.13

3.4 Correlations Between Different Measures

Spearman correlations were calculated between the most representative measures from each dimension (parameters with a significant coverage, intensity and density) according to the results obtained in PlumX. These are: Readers in Mendeley, Citations in Scopus, number of tweets in Twitter, and Abstracts Views in EBSCO. Furthermore, PDF Views and HTML Views in EBSCO have been used as an approach for downloads in EBSCO. The results for the three publication years are shown in Table 9.

Table 9. Correlations between the top measures for each dimension in PlumX and each publication year (in bold= higher than 0.4)

	Parameter	Citations: Scopus	Social media: Tweets: Twitter	Usage: abstract views: EBSCO	Usage: HTML/PDF views: EBSCO
2014	Captures: Readers: Mendeley	**0.45**	0.17	0.14	0.07
	Citations: Scopus	1	0.12	0.08	0.02
	Social Media: Tweets:Twitter	0.12	1	0.06	0.03
	Usage:Abstract Views:EBSCO	0.08	0.06	1	**0.68**
	Usage: HTML/PDF Views:EBSCO	0.02	0.03	0.68	1
2015	Captures: Readers: Mendeley	**0.60**	0.09	0.25	0.06
	Citations: Scopus	1	0.17	0.22	0.02
	Social Media: Tweets:Twitter	0.17	1	0.05	<0.01
	Usage:Abstract Views:EBSCO	0.22	0.05	1	**0.48**
	Usage: HTML/PDF Views:EBSCO	0.02	<0.01	0.48	1
2016	Captures: Readers: Mendeley	**0.46**	0.40	0.17	0.04
	Citations: Scopus	1	0.38	0.11	<0.01
	Social Media: Tweets:Twitter	0.38	1	0.12	0.02
	Usage:Abstract Views:EBSCO	0.11	0.12	1	**0.50**
	Usage: HTML/PDF Views:EBSCO	<0.01	0.02	0.50	1

The results show a very low, almost insignificant correlation between the different dimensions. However, values are increasing along with increasing measuring windows (higher correlation values for the publication year 2014 with the larger window).

Table 10. Correlations between WoS CC citation (TC and Z9) and usage counts (U1 and U2)

Year	Parameter	TC	Z9
2014	U1	0.47226475	0.46737161
2014	U2	0.45563822	0.45170766
2015	U1	0.57529206	0.57577977
2015	U2	0.48976921	0.4897413
2016	U1	0.47288015	0.47675015
2016	U2	0.26353792	0.26557909

Table 10 informs about the correlations computed in WoS between citation (times cited in WoS Core Collection or TC and times cited in the complete WoS platform or Z9) and usage counts (U1 and U2, as described in the methodology). It shows a medium correlation except for the publication year 2016 with the smallest citation window.

Finally, correlations were computed between the number of authors or the number of affiliations and the most representative measures traced in each data source (see Methodology). Concerning the number of authors, the obtained Spearman correlation values were always insignificant (less than 0.1). Concerning the number of affiliations, more significant values were only observed for captures (0.25 in 2015) and citation counts (varying from 0.2 to 0.4).

4   Conclusion and Discussion

The results of our monitoring exercise reinforce the importance of the usage metrics in order to assess the broad impact of journal articles, especially in disciplines related to the Arts & Humanities. This confirms that publications in this area are often viewed or downloaded due to the fact that they are used for other purposes (pure information, learning, teaching, etc.) apart from the 'publish or perish game' [17, 18]. A comparison with previous results for journal articles shows that the percentage of journal articles with usage data is very similar to the one reported in previous studies [3, 17]. The importance of citation data will increase with the longer citation window according to the different cited and citing half–life characteristic for each area and discipline.

However, the highest coverage or degree of data availability is provided by the number of readers in Mendeley independent of the knowledge area in agreement with previous results [19]. Almost 90% of the Austrian WoS Core Collection publications with DOIs were captured by at least by one reader in this reference manager even in the more recent years.

Concerning altmetrics [4], the percentage of Austrian WoS CC publications with social media scores is strongly increasing from ∼28% in 2014 to ∼38% in 2016, in agreement with the increasing popularity and advancement of these tools in recent years [20]. According to our results, the highest percentages of data availability in social media are reported in the Health and Life Sciences, followed very closely by the Social Sciences, where they play a significant role [20]. A low reported percentage of data availability for the Physical Sciences and the relative insignificance in the Arts & Humanities is noteworthy. In particular, the tweets intensity has even been increasing within the last years despite the reduced time window. This hints at an increasing use of social media within the scholarly community in general [21], and particularly in an increase of Twitter usage in Austria.

According to our results – very low correlation values between the measures traced in PlumX - different dimensions might provide only partial views, and they should be considered rather as complementary sources in order to reach a higher completeness of data [22]. High correlations between the same measures or metrics originating from different data sources were only reported for citations but not for usage data. Medium correlation values were reported between usage and citation counts in the database WoS Core Collection. These results are well in agreement with previous results reported by Chi and Glänzel [23]. According to these, citations and usage counts in WoS correlate significantly, especially in the Social Sciences. However, one should bear in mind that usage data and citations have different obsolescence patterns. Most articles are viewed or downloaded immediately upon their online availability. In many cases, they might reach their download maximum even before they appear in print format. In contrast – and depending on the research area, it takes a couple of years until articles receive their citation peak [3, 17]. The same authors reported that higher numbers of co-authors are not associated with higher usage counts or more citations. This hypothesis was also checked in our study and we could not find any association of the number of co-authors or co-affiliations with any of the measures considered in this study (neither in WoS Core Collection nor in PlumX).

The low correlation of usage data with citations and social media data is in agreement with de Winter's analysis for PLOS ONE articles, revealing that the number of tweets is weakly associated with the number of citations and it is only predictive of other social media activity (e.g. Mendeley and for Facebook), but not for usage data [24]. An almost non-existent correlation between EBSCO abstracts, PDF or HTML views and the number of readers in Mendeley is also an interesting fact. This hints at quite different user communities.

The aim of our study is to provide a first example of monitoring the web impact of the publication output on country level. A restriction of our study is that more than 90% of the publications with DOI retrieved in WoS were journal articles and reviews. An analysis of the web impact of the total national publication output should, of course, also consider other publication types, even if they lack a DOI. For other document types, like for example books, big differences are expected but should be feasible to process by using ISBNs as identifiers (see Torres Salinas et al. 2017) for books, or other appropriate identifiers according to the analyzed publication type.

PlumX has proven to be a very useful tool in order to monitor the broad impact of the publication output at country or institution level. Our example also strengthens the philosophy of the PlumX tool providing a cornucopia of measures grouped in different dimensions, but not providing a simple and composite indicator [11]. In doing so, the multidimensional aspect is better addressed, even if it is far from trivial in dealing with such an amalgam of different types of information retrieved from a plethora of data sources [21].

Further research is necessary to clarify the stability and reproducibility of altmetrics data, in order to get a thorough and transparent documentation of their temporal evolution and to trace and understand potential score changes. Unfortunately, PlumX (as it is common for all current tools tracing this kind of data) does not offer the possibility to select different measuring windows, as it is possible in citation indices. Therefore, temporal monitoring currently only works by archiving obtained results and later comparison at different time intervals.

New metrics should not only be used for evaluation purposes, but also in order to trace and monitor the interest and attention attracted by the publication output of a country or institution [25], and to follow their evolution in time. This could be beneficial for developing more suitable services for scientists, institutions and countries to increase their visibility on the web.

Acknowledgments. The authors thank Christina Lohr and Tina Moir from Elsevier for granted trial access to PlumX as well as Christian Gumpenberger for his help.

Appendix

See Tables 11, 12 and 13.

Table 11. Results for the publication year 2014

		Items with data scores >= 1	# Records with data scores >=1 / # WoS CC Records	Intensity (Sum)	Mean	Density (Mean available)	Median	Maximum	Standard deviation	T confidence interval (α = 0.05)
		PY= 2014 (in: 16,626; out: 16,538)								
Captures	Exports-Saves:EBSCO	9,590	57.99%	121,595	7.35	12.68	1	3,653	41.45	0.63
	Readers:Mendeley	15,299	92.51%	342,723	20.72	22.40	10	1,753	44.78	0.68
	Total Captures	15,622	94.46%	464,318	28.08	29.72	13	3,746	64.30	0.98
Citations	Clinical PubMed Guidelines	198	1.20%	302	0.02	1.53	0	10	0.22	<0.01
	Clinical DynaMed Plus	148	0.89%	206	0.01	1.39	0	11	0.17	<0.01
	Clinical NICE	8	0.05%	8	<0.01	1	0	1	0.02	<0.01
	Citation Indexes:PubMedCE	298	1.80%	838	0.05	2.81	0	26	0.55	<0.01
	Citation Indexes:Scopus	14,058	85.00%	201,533	12.19	14.34	6	2,563	37.09	0.57
	Citation Indexes:PubMed	4,921	29.76%	39,270	2.37	7.98	0	854	11.30	0.17
	Citation Indexes:RePEc	41	0.25%	416	0.03	10.15	0	66	0.80	0.01
	Citation Indexes:SciELO	3	0.02%	5	<0.01	1.67	0	2	0.02	<0.01
	Citation Indexes:CrossRef	13,515	81.72%	175,586	10.62	12.99	5	1,457	28.07	0.43
	Policy Policy Citation	1	0.01%	1	<0.01	1	0	1	<0.01	<0.01
	Total Citations	14,549	87.97%	418,165	25.29	28.74	11	3,786	71.74	1.09
Social Media	+1s:Google+	332	2.01%	1,542	0.09	4.64	0	263	2.65	0.04
	Tweets:Twitter	4,198	25.38%	38,161	2.31	9.09	0	2,481	33.60	0.51
	Shares, Likes & Comments:Faceboo	1,485	8.98%	83,666	5.06	56.34	0	27,296	222.40	3.39
	Total Social Media	4,715	28.51%	123,369	7.46	26.17	0	28,686	237.00	3.61
Mentions	Blog Mentions:Blog	339	2.05%	1,567	0.09	4.62	0	1,023	7.96	0.12
	Economics Blog Mentions:Blog	11	0.07%	14	<0.01	1.27	0	2	0.03	<0.01
	Comments:Reddit	51	0.31%	3,734	0.23	73.22	0	2,228	19.21	0.29
	Links:Wikipedia	359	2.17%	517	0.03	1.44	0	17	0.31	<0.01
	News Mentions:Blog	1	0.01%	1	<0.01	1	0	1	<0.01	<0.01
	News Mentions:News	225	1.36%	1,166	0.07	5.18	0	360	3.25	0.05
	Total Mentions	815	4.93%	6,999	0.42	8.59	0	2,239	22.20	0.34
Usage	Sample Downloads:EBSCO	1	0.01%	2	<0.01	2.00	0	2	0.02	<0.01
	Views:Figshare	26	0.16%	2,741	0.17	105.42	0	1,065	9.25	0.14
	Abstract Views:EBSCO	14,650	88.58%	2,048,497	123.87	139.83	25	35,784	547.91	8.35
	Abstract Views:SSRN	1	0.01%	196	0.01	196	0	196	1.52	0.02
	Abstract Views:DSpace	100	0.60%	2,151	0.13	21.51	0	1,210	9.58	0.15
	Abstract Views:airiti Library	5	0.03%	210	0.01	42.00	0	148	1.20	0.02
	Abstract Views:RePEc	50	0.30%	4,740	0.29	94.80	0	417	7.03	0.11
	Abstract Views:SciELO	14	0.08%	1,224	0.07	87.43	0	424	3.91	0.06
	Abstract Views:Bepress	3	0.02%	80	<0.01	26.67	0	68	0.53	<0.01
	Data Views:EBSCO	2	0.01%	2	<0.01	1	0	1	0.01	<0.01
	PDF Views:PubMedCentral	471	2.85%	98,066	5.93	208.21	0	2,500	47.58	0.73
	PDF Views:SciELO	14	0.08%	3,686	0.22	263.29	0	729	9.22	0.14
	PDF Views:PLoS	461	2.79%	421,796	25.50	914.96	0	214,171	1668.44	25.43
	PDF Views:EBSCO	838	5.07%	43,925	2.66	52.42	0	3,813	52.99	0.81
	Clicks:Bitly	617	3.73%	29,229	1.77	47.37	0	6,040	55.81	0.85
	HTML Views:PubMedCentral	471	2.85%	454,435	27.48	964.83	0	130,402	1057.90	16.12
	HTML Views:SciELO	14	0.08%	11,488	0.69	820.57	0	1,510	26.65	0.41
	HTML Views:PLoS	461	2.79%	1,691,534	102.28	3669.27	0	260,364	2255.02	34.37
	HTML Views:EBSCO	4,463	26.99%	284,700	17.21	63.79	0	23,692	236.61	3.61
	Downloads:Figshare	26	0.16%	724	0.04	27.85	0	116	1.72	0.03
	Downloads:DSpace	16	0.10%	671	0.04	41.94	0	273	2.53	0.04
	Downloads:UWA Research Reposito	6	0.04%	155	<0.01	25.83	0	55	0.60	<0.01
	Downloads:airiti Library	4	0.02%	31	<0.01	7.75	0	28	0.22	<0.01
	Downloads:D-Scholarship@Pitt	6	0.04%	704	0.04	117.33	0	142	2.26	0.03
	Downloads:RePEc	49	0.30%	1,233	0.07	25.16	0	140	1.98	0.03
	Downloads:Bepress	2	0.01%	527	0.03	263.50	0	503	3.92	0.06
	Full Text Views: ResearchSPAce	1	0.01%	54	<0.01	54	0	54	0.42	<0.01
	Link-outs:EBSCO	10,156	61.41%	187,900	11.36	18.50	1	5,033	60.58	0.92
	Total Usage	14,721	89.01%	5,290,701	319.91	359.40	32	513,918	4451.57	67.85
	Total All	16,203	97.97%	6,303,552	381.16	389.04	75	545,537	4696.25	71.58

Table 12. Results for the publication year 2015

		PY=2015 (in: 18,615; out: 18,486)							
		Items with data scores > = 1	# Records with data scores >=1 / # WoS CC Records	Intensity (Sum)	Mean	Density (Mean available)	Median	Maximum	Standard deviation	T confidence interval (α = 0.05)
Captures	Exports-Saves:EBSCO	7,312	39.55%	91,600	4.96	12.53	0	1,772	23.06	0.33
	Readers:Mendeley	16,780	90.77%	301,235	16.30	17.95	8	1,196	32.88	0.47
	Total Captures	17,017	92.05%	392,835	21.25	23.08	10	1,790	42.75	0.62
Citations	Clinical PubMed Guidelines	86	0.47%	101	<0.01	1.17	0	5	0.09	<0.01
	Clinical DynaMed Plus	157	0.85%	208	0.01	1.32	0	7	0.14	<0.01
	Clinical NICE	1	0.01%	1	<0.01	1.00	0	1	<0.01	<0.01
	Citation Indexes:Scopus	14,544	78.68%	133,910	7.24	9.21	3	1,509	21.87	0.32
	Citation Indexes:RePEc	1	0.01%	3	<0.01	3	0	3	0.02	<0.01
	Citation Indexes:PubMed	4,545	24.59%	26,737	1.45	5.88	0	788	9.59	0.14
	Citation Indexes:PubMedCE	96	0.52%	142	<0.01	1.48	0	14	0.15	<0.01
	Citation Indexes:CrossRef	14,348	77.62%	124,838	6.75	8.70	3	1,332	19.92	0.29
	Total Citations	15,367	83.13%	285,940	15.47	18.61	7	3,631	50.08	0.72
Social Media	+1s:Google+	265	1.43%	1,158	0.06	4.37	0	252	2.10	0.03
	Tweets:Twitter	5,726	30.97%	64,695	3.50	11.30	0	15,007	113.42	1.64
	Shares, Likes & Comments:Facebook	1,572	8.50%	92,378	5	58.76	0	10,676	130.53	1.88
	Total Social Media	6,128	33.15%	158,231	8.56	25.82	0	15,179	181.49	2.62
Mentions	Blog Mentions:Blog	456	2.47%	964	0.05	2.11	0	113	0.96	0.01
	Economics Blog Mentions:Blog	7	0.04%	12	<0.01	1.71	0	5	0.04	<0.01
	Comments:Reddit	70	0.38%	1,982	0.11	28.31	0	1,136	8.63	0.12
	Links:Wikipedia	344	1.86%	756	0.04	2.20	0	136	1.25	0.02
	News Mentions:News	615	3.33%	3,494	0.19	5.68	0	498	5.01	0.07
	Total Mentions	1,183	6.40%	7,208	0.39	6.09	0	1,136	10.32	0.15
Usage	Sample Downloads:EBSCO	1	0.01%	2	<0.01	2	0	2	0.01	<0.01
	Views:Figshare	10	0.05%	872	0.05	87.20	0	263	2.61	0.04
	Abstract Views:CABI	1	0.01%	3	<0.01	3	0	3	0.02	<0.01
	Abstract Views:EBSCO	15,318	82.86%	1,700,018	91.96	110.98	16	27,287	387.30	5.58
	Abstract Views:SSRN	1	0.01%	186	0.01	186	0	186	1.37	0.02
	Abstract Views:DSpace	161	0.87%	2,557	0.14	15.88	0	419	4.51	0.07
	Abstract Views:airiti Library	5	0.03%	21	<0.01	4.20	0	7	0.07	<0.01
	Abstract Views:RePEc	9	0.05%	54	<0.01	6	0	41	0.30	<0.01
	Abstract Views:SciELO	7	0.04%	494	0.03	70.57	0	97	1.41	0.02
	Abstract Views:Bepress	3	0.02%	42	<0.01	14	0	21	0.19	<0.01
	Holdings:WorldCat	5	0.03%	959	0.05	191.80	0	435	3.90	0.06
	PDF Views:PubMedCentral	428	2.32%	63,811	3.45	149.09	0	1,170	28.66	0.41
	PDF Views:SciELO	7	0.04%	2,846	0.15	406.57	0	851	9.22	0.13
	PDF Views:PLoS	414	2.24%	137,627	7.44	332.43	0	2,606	63.45	0.91
	PDF Views:EBSCO	1	0.01%	56	<0.01	56	0	56	0.41	<0.01
	Clicks:Bitly	667	3.61%	45,180	2.44	67.74	0	17,078	128.88	1.86
	HTML Views:PubMedCentral	428	2.32%	182,923	9.90	427.39	0	2,902	87.40	1.26
	HTML Views:SciELO	7	0.04%	9,084	0.49	1297.71	0	3,151	30.05	0.43
	HTML Views:PLoS	414	2.24%	799,679	43.26	1931.59	0	18,809	419.79	6.05
	HTML Views:EBSCO	4,730	25.59%	229,673	12.42	48.56	0	9,690	111.22	1.60
	Downloads:RePEc	2	0.01%	18	<0.01	9	0	17	0.13	<0.01
	Downloads:Figshare	10	0.05%	481	0.03	48.10	0	140	1.52	0.02
	Downloads:DSpace	28	0.15%	693	0.04	24.75	0	154	1.60	0.02
	Downloads:UWA Research Repository	11	0.06%	264	0.01	24	0	63	0.74	0.01
	Downloads:airiti Library	5	0.03%	6	<0.01	1.20	0	2	0.02	<0.01
	Downloads:D-Scholarship@Pitt	6	0.03%	231	0.01	38.50	0	118	0.95	0.01
	Downloads:PhilSci-Archive	2	0.01%	193	0.01	96.50	0	97	1	0.01
	Downloads:Bepress	3	0.02%	77	<0.01	25.67	0	45	0.37	<0.01
	FT Views:Nottingham Trent Univ-IRep	2	0.01%	15	<0.01	7.50	0	10	0.08	<0.01
	Link-outs:EBSCO	10,614	57.42%	195,771	10.59	18.44	1	4,215	57.38	0.83
	Total Usage	15,430	83.47%	3,373,836	182.51	218.65	21	28,042	781.01	11.26
	Total All	17,981	97.27%	4,218,050	228.18	234.58	51	29,836	843.67	12.16

Table 13. Results for the publication year 2016

		Items with data scores >= 1	# Records with data scores >=1 / # WoS CC Records	Intensity (Sum)	Mean	Density (Mean available)	Median	Maximum	Standard deviation	T confidence interval (α = 0.05)
				PY= 2016 (in: 20,631; out: 20,452)						
Captures	Exports-Saves:EBSCO	6,153	30.09%	56,584	2.77	9.20	0	988	13.47	0.18
	Readers:Mendeley	17,857	87.31%	240,296	11.75	13.46	5	834	24.17	0.33
	Total Captures	18,116	88.58%	296,880	14.52	16.39	6	1,175	29.83	0.41
Citations	Clinical PubMed Guidelines	34	0.17%	37	<0.01	1.09	0	2	0.05	<0.01
	Clinical DynaMed Plus	110	0.54%	187	<0.01	1.70	0	44	0.33	<0.01
	Citation Indexes:Scopus	12,863	62.89%	64,197	3.14	4.99	1	527	9.31	0.13
	Citation Indexes:PubMed	3,158	15.44%	11,435	0.56	3.62	0	378	4.19	0.06
	Citation Indexes:PubMedCE	5	0.02%	5	<0.01	1.00	0	1	0.02	<0.01
	Citation Indexes:CrossRef	12,826	62.71%	61,783	3.02	4.82	1	505	9.19	0.13
	Total Citations	14,175	69.31%	137,644	6.73	9.71	2	1,301	21.46	0.29
Social Media	+1s:Google+	106	0.52%	328	0.02	3.09	0	78	0.68	<0.01
	Tweets:Twitter	7,503	36.69%	73,031	3.57	9.73	0	1,005	24.26	0.33
	Shares, Likes & Comments:Facebook	1,525	7.46%	79,026	3.86	51.82	0	3,867	60.43	0.83
	Total Social Media	7,824	38.26%	152,385	7.45	19.48	0	4,151	73.92	<0.01
Mentions	Blog Mentions:Blog	500	2.44%	1,147	0.06	2.29	0	45	0.67	<0.01
	Economics Blog Mentions:Blog	8	0.04%	9	<0.01	1.13	0	2	0.02	<0.01
	Comments:Reddit	74	0.36%	834	0.04	11.27	0	268	2.05	0.03
	Links:Wikipedia	250	1.22%	383	0.02	1.53	0	31	0.30	<0.01
	News Mentions:News	946	4.63%	3,940	0.19	4.16	0	168	2.38	0.03
	Total Mentions	1,425	6.97%	6,313	0.31	4.43	0	269	3.47	0.05
Usage	Sample Downloads:EBSCO	1	<0.01%	1	<0.01	1.00	0	1	<0.01	<0.01
	Views:Figshare	5	0.02%	72	<0.01	14.40	0	30	0.29	<0.01
	Abstract Views:SSRN	1	<0.01%	109	<0.01	109.00	0	109	0.76	0.01
	Abstract Views:DSpace	131	0.64%	2,886	0.14	22.03	0	260	3.17	0.04
	Abstract Views:EBSCO	16,271	79.56%	1,369,298	66.95	84.16	10	13,911	274.81	3.77
	Abstract Views:RePEc	7	0.03%	37	<0.01	5.29	0	14	0.13	<0.01
	Abstract Views:SciELO	16	0.08%	522	0.03	32.63	0	85	1.14	0.02
	Abstract Views:Bepress	3	0.01%	88	<0.01	29.33	0	73	0.52	<0.01
	Holdings:WorldCat	4	0.02%	634	0.03	158.50	0	427	3.10	0.04
	PDF Views:PubMedCentral	377	1.84%	31,777	1.55	84.29	0	574	15.57	0.21
	PDF Views:SciELO	16	0.08%	3,189	0.16	199.31	0	1,411	10.88	0.15
	PDF Views:PLoS	369	1.80%	109,175	5.34	295.87	0	3,703	56.61	0.78
	PDF Views:EBSCO	3	0.01%	3,156	0.15	1052.00	0	2,717	19.22	0.26
	Clicks:Bitly	868	4.24%	27,297	1.33	31.45	0	3,968	41.25	0.57
	HTML Views:PubMedCentral	378	1.85%	89,560	4.38	236.93	0	1,574	43.11	0.59
	HTML Views:SciELO	16	0.08%	3,848	0.19	240.50	0	661	8.25	0.11
	HTML Views:PLoS	369	1.80%	604,071	29.54	1637.05	0	16,481	344.81	4.73
	HTML Views:EBSCO	2,874	14.05%	158,635	7.76	55.20	0	11,536	119.80	1.64
	Downloads:PhilSci-Archive	1	<0.01%	110	<0.01	110.00	0	110	0.77	0.01
	Downloads:Figshare	6	0.03%	45	<0.01	7.50	0	18	0.17	<0.01
	Downloads:DSpace	17	0.08%	559	0.03	32.88	0	128	1.32	0.02
	Downloads:UWA Research Repository	4	0.02%	30	<0.01	7.50	0	12	0.12	<0.01
	Downloads:D-Scholarship@Pitt	5	0.02%	197	<0.01	39.40	0	71	0.72	<0.01
	Downloads:RePEc	4	0.02%	11	<0.01	2.75	0	6	0.05	<0.01
	Downloads:Bepress	3	0.01%	72	<0.01	24.00	0	53	0.39	<0.01
	FT Views:Journal World-Systems Res	3	0.01%	1,722	0.08	574.00	0	671	7.04	0.10
	FT Views:Nottingham Trent Univ- IRe	2	<0.01%	17	<0.01	8.50	0	12	0.09	<0.01
	Link-outs:EBSCO	10,007	48.93%	138,961	6.79	13.89	0	1,926	35.55	0.49
	Total Usage	16,449	80.43%	2,546,079	124.49	154.79	12	23,040	594.67	8.15
	Total All	19,596	95.81%	3,139,301	153.50	160.20	31	23,177	624.16	8.55

References

1. Gorraiz, J., Wieland, M., Gumpenberger, C.: To be visible, or not to be, that is the question. Int. J. Soc. Sci. Humanity **7**(7), 467–471 (2017)
2. Kraemer, A.: Ensuring consistent usage statistics, part 2: working with use data for electronic journals. Ser. Librarian **50**(1/2), 163–172 (2006)
3. Gorraiz, J., Gumpenberger, C., Schloegl, C.: Usage versus citation behaviours in four subject areas. Scientometrics **101**(2), 1077–1095 (2014)
4. Glänzel, W., Gorraiz, J.: Usage metrics versus altmetrics: confusing terminology? Scientometrics **102**, 2161–2164 (2015)
5. Procter, R., Williams, R., Stewart, J., Poschen, M., Snee, H., Voss, A., Asgari-Targhi, M.: Adoption and use of Web 2.0 in scholarly communications. Philos. Trans. Ser. A Math. Phys. Eng. Sci. **368**(1926), 4039–4056 (2010)
6. Kietzmann, J.H., Hermkens, K., McCarthy, I.P., Silvestre, B.S.: Social media? Get serious! Understanding the functional building blocks of social media. Bus. Horiz. **54**(3), 241–251 (2011)
7. Wouters, P., Costas, R.: Users, narcissism and control – Tracking the impact of scholarly publications in the 21st century (2012). http://research-acumen.eu/wp-content/uploads/Users-narcissism-and-control.pdf
8. Haustein, S., Peters, I., Bar-Ilan, J., Priem, J., Shema, H., Terliesner, J.: Coverage and adoption of altmetrics sources in the bibliometric community. In: Gorraiz, J., Schiebel, E., Gumpenberger, C., Hörlesberger, M., Moed, H. (eds.) Proceedings of the 14th International Society of Scientometrics and Informetrics, vol. 2, pp. 1–12 (2013)
9. Gorraiz, J., Gumpenberger, C.: Bibliometric practices and activities at the University of Vienna. Libr. Manag. **33**(3), 174–183 (2012)
10. Bornmann, L.: Do altmetrics point to the broader impact of research? An overview of benefits and disadvantages of altmetrics. J. Informetrics **8**(4), 895–903 (2014)
11. Gorraiz, J., Melero-Fuentes, D., Gumpenberger, C., Valderrama-Zurián, J.C.: Availability of digital object identifiers (DOIs) in Web of Science and Scopus. J. Informetrics **10**(1), 98–109 (2016)
12. Torres-Salinas, D., Gumpenberger, C., Gorraiz, J.: PlumX as a potential tool to assess the macroscopic multidimensional impact of books. Front. Res. Metr. Anal. (2017). https://doi.org/10.3389/frma.2017.00005
13. Haustein, S., Costas, R., Larivière, V.: Characterizing social media metrics of scholarly papers: the effect of document properties and collaboration patterns. PLoS One **10**(3), e0120495 (2015). https://doi.org/10.1371/journal.pone.0120495
14. Peters, I., Kraker, P., Lex, E., Gumpenberger, C., Gorraiz, J.: Zenodo in the spotlight of old and new metrics. Front. Res. Metr. Anal. (2017, in press)
15. Archambault, E., Campbell, D., Gingras, Y., Lariviere, V.: Comparing of science bibliometric statistics obtained from the Web and Scopus. J. Am. Assoc. Inf. Sci. Technol. **60**(7), 1320–1326 (2009). https://doi.org/10.1002/asi.21062
16. Gorraiz, J., Schlögl, C.: A bibliometric analysis of pharmacology and pharmacy journals: Scopus versus Web of Science. J. Inf. Sci. **34**, 715–725 (2008)
17. Schlögl, C., Gorraiz, J.: Comparison of citation and usage indicators: the case of oncology journals. Scientometrics **82**, 567–580 (2010)
18. Bollen, J., Van de Sompel, H.: Usage impact factor: the effects of sample characteristics on usage-based impact metrics. J. Am. Soc. Inform. Sci. Technol. **59**(1), 136–149 (2008)

19. Zahedi, Z., Costas, R., Wouters, P.: How well developed are Altmetrics? Cross disciplinary analysis of the presence of 'alternative metrics' in scientific publications (RIP). In: Gorraiz, J., Schiebel, E., Gumpenberger, C., Hörlesberger, M., Moed, H. (eds.) Proceedings of the 14th International Society of Scientometrics and Informetrics Conference, vol. 1, pp. 876–884 (2013)
20. Sugimoto, C.R., Work, S., Lariviere, V., Haustein, S.: Scholarly use of social media and altmetrics: a review of the literature. J. Am. Assoc. Inf. Sci. Technol. **68**(9), 2037–2062 (2017)
21. Robinson-Garcia, N., Costas, R., Isett, K., Melkers, J., Hicks, D.: The unbearable emptiness of tweeting—about journal articles. PLoS One **12**(8), e0183551 (2017). https://doi.org/10.1371/journal.pone.0183551
22. Gorraiz, J., Glänzel, W., Gumpenberger, C.: The ecstasy and agony of the altmetric score. Scientometrics **108**(2), 977–982 (2016)
23. Chi, P., Glänzel, W.: An empirical investigation of the associations among usage, scientific collaboration and citation impact. Scientometrics **112**(1), 403–412 (2017)
24. De Winter, J.C.F.: The relationship between tweets, citations, and article views for PLOS ONE articles. Scientometrics **102**(2), 1773–1779 (2015)
25. Alhoori, H., Furuta, R., Tabet, M., Samaka, M., Fox, E.A.: Altmetrics for country-level research assessment. In: Tuamsuk, K., Jatowt, A., Rasmussen, E. (eds.) ICADL 2014. LNCS, vol. 8839, pp. 59–64. Springer, Cham (2014). https://doi.org/10.1007/978-3-319-12823-8_7

New Dialog, New Services with Altmetrics: Lingnan University Library Experience

Sheila Cheung$^{(\boxtimes)}$ ⓘ, Cindy Kot ⓘ, and Kammy Chan ⓘ

Fong Sum Wood Library, Lingnan University,
8 Castle Peak Road, Tuen Mun, Hong Kong
sheila@LN.edu.hk

Abstract. This paper chronicles Lingnan University library's approach to integrating altmetrics into its institutional repository, triggering new dialog with Lingnan's researchers and to re-energize their attention on the scholarly communication services with new perspective.

Keywords: IR-Altmetrics integration · Scholarly communication
Research impact

1 Introduction

Altmetrics poses new challenges and opportunities for academic libraries. The Association of College and Research Libraries (ACRL) identified altmetrics as one of the top trends in academic libraries in two consecutive reports of the years 2014 and 2016 [1, 2]. Altmetrics is considered as a natural extension of what academic libraries excelled in, which is presenting various possibilities to build up new roles for changing the land-scape of scholarly communication [3]. This paper chronicles Lingnan University library's approach to integrating altmetrics into its institutional repository (hereafter referred to as IR) for triggering new dialog with our researchers and to re-energize their attention on our scholarly communication services with new perspective.

2 Background and Literature Review

Lingnan University, currently the only liberal arts university in Hong Kong, was originally founded in Guangzhou (Canton), China in 1888, however, suspended in 1952. Re-established in Hong Kong by its alumni as Lingnan College in 1967, it became part of the public-funded tertiary system in 1991 and accredited to become Lingnan University in 1999. As of 2017/2018 academic year, Lingnan University consists of 16 academic departments functioning under three faculties, i.e. Arts, Business and Social Sciences, plus other 13 research institutes or centers. There are a total of 188 full-time academic staff and over 3,000 FTE students.

Lingnan Library began its operation since 1968, currently holding more than 530,000 physical items, accompanied with a wide range of electronic collection.

© Springer Nature Singapore Pte Ltd. 2018
M. Erdt et al. (Eds.): AROSIM 2018, CCIS 856, pp. 63–71, 2018.
https://doi.org/10.1007/978-981-13-1053-9_5

The Scholarly Communication Team was formed in April 2012 alongside with the launch of the first IR platform, *Digital Commons@Lingnan University* (thereafter referred to as DC@Lingnan), committed to raise awareness of Open Access and to support Lingnan's researchers in sharing their intellectual life with a global audience. As of September 2017, more than 200 content series were arranged in DC@Lingnan, containing more than 11,600 records, all with enriched metadata reflecting both bibliographic information and other essential fields to facilitate backend repository management.

During the first two years, DC@Lingnan acquired content primarily by migrating existing items from library's internal archives, but soon encountered difficulties to sustain its growth due to low researcher engagement in supplying new content. Despite our efforts of advertising various benefits that DC@Lingnan could bring, e.g. OA advantage, increased visibility and discoverability, citation advantage, etc., all these have never been transformed into real incentives to get researchers to contribute content. Research Libraries UK's study [4] found many researchers were unaware of the benefit of IR and considered depositing content as a "burden". Ironically, many researchers have indeed created their personal profiles on multiple academic networking platforms, e.g. Google Scholar or ResearchGate, having their scholarly outputs neatly presented. DC@Lingnan, being the platform offered by their respective institute carrying similar functions, was not even considered as an alternative. Obviously, the value of DC@Lingnan has never been fully seen by researchers.

Bankier and Smith [5] related such detachments with libraries' failure in framing IR to resonate with faculty. Generating appropriate IR resonance to attract attention, however, involved a steep learning curve. Our initial DC@Lingnan outreach hinted to us that metrics may be the key. Presenting IR usage was the only topic among others that could trigger extended dialog with researchers.

Metrics touch the nerve across academia, as it may bring impact over appraisal, tenure or soliciting research funds, etc. Traditional bibliometrics based upon citation counts have long been the dominating benchmark to reflect research impact. But metrics are getting more diverse and complex due to rapid technological development. Priem and Hemminger [6] raised the term "Scientometrics 2.0" in 2010, and contended that Web 2.0 environment would facilitate more active participation across scholarly communication, leading to faster, broader and more comprehensive ways to measure impact with metrics drawn from the social web. The Altmetrics Manifesto was later published advocating the creation and study of new metrics based on the social web [7].

Altmetrics, abbreviated from "Alternative Metrics" provide various non-traditional indicators, e.g., download or view usage, comments, shares, captures, etc., visualizing whether and how impacts are generated through different online media. For offering any service based upon metrics, Armbruster [8] suggested it would require a database of which IR, containing carefully curated metadata by library, was seen as a reliable source to begin with. With the growing attention of altmetrics, Lingnan library believed that by adding altmetrics elements to DC@Lingnan, it would create new opportunities to alleviate the IR dilemma. Our goal is to utilize appropriate altmetrics tools to create IR resonance, intensify dialogue with different stakeholders, and to generate more direct and indirect engagement with our researchers.

3   Initial Altmetrics Adoption – Full-Text Download Usage

Initial altmetrics initiatives began with consolidating download usage reports by using the default usage utilities of DC@Lingnan. The reports showing cumulative download counts of each publication were sent to individual departments. DC@Lingnan had attracted good readership since it was first launched. The usage report speaks for the benefits and helped transform stakeholders' perspective to our services.

The Office of Service-Learning (hereafter referred as OSL) was among the first department to actively engage with DC@Lingnan after reading the download usage report. OSL collaborates with faculties, offering learning opportunities for over 600 students every academic year to practice their knowledge by serving the community. Many scholarly and creative works were therefore generated to document the learning and research outcomes covering activity reports, conference presentations, multimedia materials, etc. OSL used to showcase these outputs on their departmental webpage, displaying only a simple list of titles with hyperlinks leading to instant full-text access.

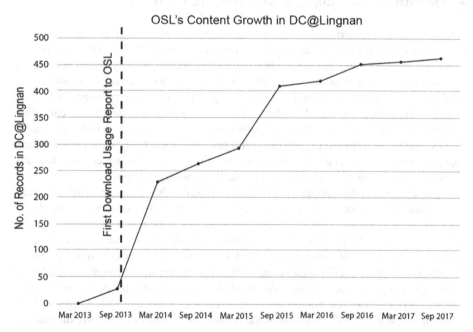

Fig. 1. OSL's content growth in DC@Lingnan (March 2013–September 2017)

Lingnan library approached OSL in early 2013, with a proposal to host their content on DC@Lingnan, aiming at disseminating their content to wider audiences. The initial response from OSL was impassive, though they raised no objection in uploading a total of 20 documents comprising their newsletters and annual reports on DC@Lingnan, as a trial. OSL's impassiveness remained until they were presented with the first usage report showing 800+ download counts recorded from those 20 uploaded documents over a six month timeframe. OSL had never been able to obtain such usage from their departmental webpage previously. The 800+ downloads from DC@Lingnan captured their attention, considering them as tangible evidence that OSL outputs were attracting readership and attention. OSL immediately realized the benefits of hosting their content on DC@Lingnan, and later began to migrate more than 450 other documents to DC@Lingnan by September 2017 (Fig. 1).

During the process, OSL and the library communicated closely regarding the migration procedure, and explored other new content that could be uploaded, e.g., hosting conference series on DC@Lingnan. A mutual trust was established where OSL considered DC@Lingnan as a reliable platform to disseminate their intellectual outputs. Their staff were also trained to get engaged with unmediated content deposit and to upload new content on a regular basis. Similar reactions were later observed from a few other academic departments, though not as active as OSL, in uploading some new content sources to help sustain the content growth of DC@Lingnan.

4 Incentivize IR Adoption with Research Impact

Lingnan University currently does not mandate content deposit to DC@Lingnan. Researchers' incentives to contribute their content are inevitably low. OSL's responses to download usage enlightened the library to realize incentives may not be necessarily needed to be tagged with mandate. Studies show that metrics could be one of the catalysts to incentivize IR adoption [9] and to generate new or continued usage [10]. With such a perspective, the library began to explore other solutions with the capability to aggregate and analyze more comprehensive metrics beyond download counts. PlumX[1], being an altmetrics aggregator with proven compatibility with DC@Lingnan appeared just in time, enabling the library to address the growing concern towards research impact with new services and outreach across the campus.

Research impact has attracted extensive attention within Lingnan in recent years. Apart from its influence over university ranking, as a publicly-funded university, Lingnan's research quality is under scrutiny through the Research Assessment Exercise (RAE) implemented by the University Grant Committee (UGC) of HKSAR Government[2]. The upcoming RAE will commence in 2020. UGC has stated explicitly in the proposed framework where "impact" would be one of the major assessment elements to be adopted in the upcoming exercise. As the RAE result is a crucial indicator of the university's research performance and competitiveness in funding allocation, this triggered huge concerns across the board towards both the quantity and quality of scholarly outputs. By bonding altmetrics with DC@Lingnan, the library made the first attempt to consolidate research impact data of Lingnan's research community, enabling the university's management to develop data-informed decision infrastructure for research management and development.

---

[1] https://plumanalytics.com/integrate/embed-metrics/.

[2] http://www.ugc.edu.hk/eng/ugc/activity/research/rae.html.

5 IR-Altmetrics Integration: DC@Lingnan with PlumX

Lin and Fenner [11] described research impact as a spectrum, with "usage" on one end of it. A series of activities would then follow, e.g. saves, recommendations, discussions, etc., finally leading to another end for having "citations" to appear. Konkiel and Scherer [12] suggested altmetrics could indicate the details of these activities taken place along the spectrum, informing researchers with the impact of their scholarly works. By leveraging the metadata in DC@Lingnan, PlumX has been used to capture these activities and convert them into altmetrics according to the level of engagement into the following five categories namely usage, captures, mentions, social media and citations[3].

Every valid record in DC@Lingnan could have their own widget generated, showing article-level altmetrics which can be further clustered into different groups. Clusterability was considered as the fundamental value of having altmetrics [13]. PlumX derived clusterability by mirroring the content hierarchy and metadata from DC@Lingnan. This however posed new problems during initial integration. DC@Lingnan was first launched to promote open access and information dissemination. The handling of staff scholarly outputs was accustomed with a simple data structure to reduce administrative overhead for a small working team. All outputs were processed in one single series, carrying mainly basic bibliographic metadata, which would be further fed into individual researcher profiles, as a showcase of their scholarly achievement (Fig. 2).

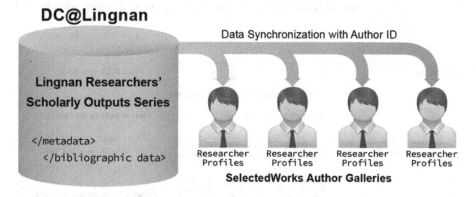

Fig. 2. Sample data structure in DC@Lingnan (before altmetrics)

Such simple structures in DC@Lingnan limited data clusterability, making it impossible to present granular altmetrics. To derive the anticipated clusterability, the library implemented a major metadata enhancement and hierarchical restructuring exercise against on the staff scholarly outputs series. Lin and Fenner [11] described this as a necessary

---

[3] PlumX Altmetrics Categories: https://plumanalytics.com/learn/about-metrics/.

"harmonization" process, involving major content grouping reconstruction and myriad organization. The process involved revamping the metadata schema of the staff publication series in DC@Lingnan with new or enhanced attributes. The content hierarchy was also restructured with extra layers embedded, so that data could be scaled to fit into various new sub-series via those enhanced attributes for each individual faculties and departments. Table 1 highlights few key enhanced attributes and their anticipated functions to facilitate altmetrics clusterability.

Table 1. Attributes and their anticipated functions to facilitate altmetrics clusterability

New/Enhanced Fields	Functions
Researchers' Affiliated Departments	Indicate publication status and research impact by Department
Researchers' Employment Status @ Lingnan	Indicate Active staff scholarly outputs, indicate the current scholarly publishing capability and impact of Lingnan

The harmonization of DC@Lingnan with PlumX took more than 6 months until the overall data synchronization between the two platforms worked satisfactorily. The structure of staff publication series became more sophisticated and accompanied with more scalable filters. While the library can maintain the existing data handling practice within one single series, the newly derived structure drives much better clusterability to support altmetrics, resulting in generating multiple levels of altmetrics dashboards. Figure 3 illustrates the different nature of altmetrics generated under simple and enhanced structures.

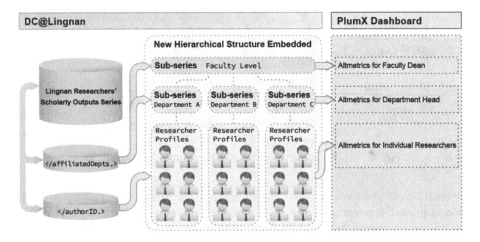

Fig. 3. Altmetrics clusterability with DC@Lingnan-PlumX integration

6   IR Resonance with Altmetrics

Empowered with PlumX, DC@Lingnan is leveraged to go beyond discovery and dissemination. Altmetrics data empowered the library to be more capable to address the needs of research impact across the campus. The library began outreach with altmetrics since mid-2016. The interaction came with diverse dialogues and responses.

6.1   Resonance at Managerial Level

Lingnan University put forth in its 2016–2022 Strategic Plan[4], a strong emphasis on producing research with impact. With this mission ahead, the university's managerial group is increasingly looking into research impact data to facilitate a data-informed decision making process. The altmetrics dashboards fit in well to fulfill such needs by providing a panoramic view of research impact of Lingnan research community, thereby offering real-time insights, not only on the productivity, but also on how research impacts were generated with different types of content across different disciplines. These dashboards were later chosen as the pinned items for regular posting on the University's Information Hub since early 2017, a platform designed to support key managerial staff to execute daily operations, strategic planning or decision-making on research support and development.

It is yet to see how altmetrics would be getting more influence against on the actual decision making process. Direct comments from a few department heads were received, indicating that altmetrics data looked useful at both departmental and individual level, and they have requested the library to prepare more guiding materials to help researchers in understanding more about the concepts. All these were positive resonances from the managerial group towards the usefulness of these altmetrics dashboards. This further strengthened the library's role with new service perspective to support research impact.

6.2   Resonance at Researchers Level

PlumX clusters altmetrics down to researcher-level by mirroring the author-galleries built within DC@Lingnan. This feature attracted attention from several individual researchers. Studies suggested that young scholars had greater need to demonstrate immediate impact to advance their academic career [14, 15]. Similar situations happened at Lingnan University where researchers with the job rank of an Assistant Professor grade or below, tended to have more obvious reactions to altmetrics. These researchers have a good understanding of bibliometrics. Altmetrics in contrast were new to them and caught their attention. They were particularly pleased with having altmetrics such as downloads, mentions or tweets recorded for their newly published works. Their interest in these numbers generated new dialog, providing us with the much needed opportunities to introduce our service rationale. Instead of emphasizing depositing content, the focus has switched to let researchers witness the diverse efforts

---

[4] http://ln.edu.hk/about-lu/strategic-plan/index.html.

the library had made to promote and manage their content. This helped in creating a good sense of "service relevancy" which Delasalle [16] considered as a crucial factor to get users engaged. A few researchers approached the library to proactively upload their new content onto DC@Lingnan, and to showcase their works in their own gallery and altmetrics dashboard. One assistant professor explicitly indicated she would include the altmetrics dashboard as a support reference in her appraisal. She also gave the following compliment, appearing as the best resonance and highlights the potential impact that altmetrics could bring to support research community at Lingnan University in the future: "…[Altmetrics] is going to provide useful information to help researchers strategize our research in alignment with the goals of [University's] strategic plan concerning research in the coming years."

6.3   Other Resonances

The core objective of having altmetrics was to incentivize IR adoption. The resulting resonance indeed went beyond this objective. Empowered with PlumX, DC@Lingnan gained more visibility and credibility across campus. The library was tagged by senior management to work together with the university's Office of Communication and Public Affairs (OCPA) for a publicity project to feature Lingnan's researchers with their impact story. The library was responsible to identify researchers with necessary metrics data and information for OCPA to develop a series of stories showing the impact and strength of Lingnan research community. With PlumX, we were able to identify interesting storylines based on altmetrics, with a research paper authored by a young professor being tweeted more than 1,000 times, and various book publications authored by professors from the Faculty of Arts, carrying several hundred WorldCat holdings. These indicators went beyond traditional citation counts and the research impact stories were more multifaceted. Most importantly, the University also uses author galleries as the official reference showcasing full publication list of the staff concerned. This was also a sign of recognition that the publication dataset handled by the library was a trustworthy source. The project offered the team a rare opportunity to collaborate with colleagues outside academic departments, and demonstrated our strength and expertise in promoting our scholarly activities for Lingnan University.

7   Conclusion

Altmetrics leverage IR capability to go beyond a repository. The bonding of DC@Lingnan with PlumX demonstrated various potentials to re-energize the traditional repository service to strike a better chord with different stakeholders. Though the preliminary resonance is yet to be further materialized for generating wider impact across campus, IR-Altmetrics integration created a good momentum to solidify the role of the library to advocate and lead the stewardship of research impact metrics, thereby forming a new service perspective to support the university's mission.

References

1. ACRL Research Planning and Review Committee: Top trends in academic libraries: a review of the trends and issues affecting academic libraries in higher education. Coll. Res. Libr. News **75**(6), 294–302 (2014)
2. ACRL Research Planning and Review Committee: Top trends in academic libraries: a review of the trends and issues affecting academic libraries in higher education. Coll. Res. Libr. News **77**(6), 274–281 (2016)
3. Sutton, S.W.: Altmetrics: what good are they to academic libraries? Kansas Libr. Assoc. Coll. & Univ. Libr. Sect. Proc. **4**(2) (2014). https://doi.org/10.4148/2160-942x.1041
4. Research Libraries UK (RLUK): Repositories increase the visibility of the institution and raise its research profile. In: RLUK & RIN. The Value of Libraries for Research and Researchers, pp. 34–37. Research Information Network, London (2011)
5. Bankier, J., Smith, C.: Digital repositories at a crossroads: achieving sustainable success through campus-wide engagement. In: VALA2010 Conference Proceedings (2010). https://works.bepress.com/jean_gabriel_bankier/8/
6. Priem, J., Hemminger, M.: Scientometrics 2.0: towards new metrics of scholarly impact on the social web. First Monday **15**(7) (2010). http://firstmonday.org/article/view/2874/2570
7. Priem, J., Taraborelli, D., Groth, P., Neylon, C.: Altmetrics: a manifesto (2010). http://altmetrics.org/manifesto/
8. Armbruster, C.: Access, usage and citation metrics: what function for digital libraries and repositories in research evaluation? SSRN (2008). https://doi.org/10.2139/ssrn.1088453
9. Roemer, R.C., Borchardt, R.: Impact and the role of Librarians. In: Roemer, R.C., Borchardt, R. (eds.) Meaningful Metrics: A 21st Century Librarian's Guide to Bibliometrics, Altmetrics, and Research Impact, pp. 209–231. Association of College and Research Librarians, Chicago (2015)
10. Bruns, T., Inefuku, H.W.: Purposeful metrics: matching institutional repository metrics to purpose and audience. In: Callicott, B.B., Scherer, D., Wesolek, A. (eds.) Making Institutional Repositories Works, pp. 213–234. Purdue University Press, West Lafayette (2016)
11. Lin, J., Fenner, M.: Altmetrics in evolution: defining and redefining the ontology of article-level metrics. Inf. Stand. Q. **25**(2), 20–26 (2013)
12. Konkiel, S., Schere, D.: New opportunities for repositories in the age of altmetrics. Bull. Assoc. Inf. Sci. Technol. **39**(4), 22–26 (2013)
13. Galligan, F., Dyas-Correia, S.: Altmetrics: rethinking the way we measure. Serials Rev. **39**, 56–61 (2013)
14. Kramer, B., Bosman, J.: Who is using altmetrics tools? In: 3:AM Bucharest 2016 (The Altmetrics Conference) (2016). http://altmetricsconference.com/who-is-using-altmetrics-tools/
15. Tattersall, A.: The connected academic: implementing altmetrics with your organization. In: Tattersall, A. (ed.) Altmetrics: a practical guide for librarians, researchers and academics, pp. 137–161. Facet, London (2016)
16. Delasalle, J.: Research evaluation: bibliometrics and the librarian. SCONUL Focus **53**, 15–19 (2011)

How Do Scholars Evaluate and Promote Research Outputs? An NTU Case Study

Han Zheng$^{(\boxtimes)}$ ⓘ, Mojisola Erdt ⓘ, and Yin-Leng Theng

Centre for Healthy and Sustainable Cities (CHESS),
Wee Kim Wee School of Communication and Information,
Nanyang Technological University, Singapore, Singapore
{zhenghan, Mojisola.Erdt, TYLTheng}@ntu.edu.sg

Abstract. Academic scholars have begun to use diverse social media tools to disseminate and evaluate research outputs. Altmetrics which are indices based on social media, have emerged in recent years and have received attention from scholars from a variety of research areas. This case study aims to investigate how researchers at Nanyang Technological University (NTU) promote their research outputs and how they conduct research evaluation. We conducted semi-structured interviews with eighteen NTU researchers. Results from the study show that NTU researchers still prefer traditional metrics such as citation count, journal impact factor (JIF) and h-index for their research evaluation. They find these metrics are essential determinants to identify the quality of a research paper. In terms of altmetrics, most NTU researchers interviewed are aware of them, but have not yet used them in research evaluation. Furthermore, we found that the main methods used to promote research outputs are attending academic conferences and publishing papers in journals. However, NTU researchers also embark on using social networking sites to disseminate new findings. The results of this study contribute to the previous literature to help understand how social media and altmetrics are influencing the traditional methods of promoting and evaluating research outputs in academia.

Keywords: Social media · Traditional metrics · Altmetrics
Research evaluation

1 Introduction

Traditionally, scholars use a handful of methods to promote their research outputs including attending academic conference, publishing papers in academic journals, networking with other researchers in research seminars etc. However, these strategies may not be applicable for all scholars due to time and resource constraints [1]. Social media tools such as blogs, wikis and social networking sites have become increasingly popular in our daily lives, with the main purpose of facilitating communication [2]. Scholars have also realized the benefits of social media and have begun to adopt diverse social media tools [3]. As such, using social media could become an effective strategy for promoting research outputs within or outside academia, regardless of geographical and temporal distance [4].

© Springer Nature Singapore Pte Ltd. 2018
M. Erdt et al. (Eds.): AROSIM 2018, CCIS 856, pp. 72–80, 2018.
https://doi.org/10.1007/978-981-13-1053-9_6

Furthermore, in the past, research outputs were evaluated by a variety of metrics such as citation counts, journal impact factor (JIF), h-index, etc. These metrics have been used by scholars across different academic disciplines for decades and could be said to be significant indicators of the impact of research works [5]. Some limitations, however still exist, for example, a paper may get its first citation only after a few years of publication [1]. In recent years, owing to the popularity of social media among scholars, altmetrics which are indices based on social media, have been emerging in academia [5]. Altmetrics can measure the outreach of research works to a wider audience in a fast and timely manner. However, there are still barriers to using altmetrics in academia, partly due to the concerns about the quality of altmetrics for research evaluation. Several studies have investigated the relationship between traditional metrics and altmetrics [6]. However, how researchers promote their research works and how they use metrics for research evaluation have not yet been investigated using a qualitative approach.

This interview case study seeks to investigate how researchers at Nanyang Technological University (NTU), promote research outputs and how they conduct research evaluation. The results from this study could help us understand the extent to which NTU researchers are using social media tools in their professional work and how they perceive the effectiveness of their current methods of promoting and evaluating research outputs. The following two research questions are addressed in this paper:

- **RQ1:** How do NTU researchers promote their research outputs?
- **RQ2:** How do NTU researchers evaluate research outputs?

2 Literature Review

Traditionally, on the article level, research evaluation in academia has been conducted by counting the number of citations of published articles [7]. On the journal level, the JIF is a popular research evaluation metric in academia [8]. Citations however only capture the academic audience of research works [9], and its turnover is slow [10]. In addition, the JIF is limited as it cannot be used for interdisciplinary comparisons [8]. As a complement to traditional metrics, alternative metrics based on social media, often referred to as altmetrics which could provide a more holistic approach to evaluating research outputs [11]. Altmetrics could be applied to evaluate research, both on an article level, as well as on a journal level [12].

Diverse social media tools have been used to different extents by scholars in academia [13]. Researchers stay up to date with the latest findings in their fields by reading other researcher's blogs and social media profiles [14]. The usage of social media however varies across academic disciplines [15]. Lack of time [14] and privacy concerns [16] are some hindrances to using social media in academia. Not all researchers use social media in their professional lives, and it seems easier to manipulate altmetrics than it is to manipulate traditional metrics [17]. Moreover, the use of social media and altmetrics is not widely recognized by most research institutions as part of their promotion review process. However, this might change in future, with the increasing use of these new tools in the research workflow, universities' promotion policies are likely to be adjusted accordingly [18]. But although social media and altmetrics are increasingly being used in academia [13], little is yet known about scholars' strategies to promote their research findings and how scholars perform research evaluation.

3 Methodology

To address the research questions, semi-structured interviews were conducted at Nanyang Technological University (NTU) Singapore, to capture the thoughts and perspectives of researchers. In total, 18 NTU researchers were invited for the interviews. The interviews were conducted between March and June in 2017. A total of 16 interview questions were directly asked to discover the behaviors and perceptions of NTU scholars in using different metrics for their research evaluation, as well as their methods to promote their research. Participants were recruited via email invitation. Each interview lasted between 10 to 20 min. All interviews were recorded with a digital recorder, and then manually transcribed. Participants were mainly from China, Singapore and India. All interviews were held in English. Personal information relating to the interviewees was kept strictly confidential. NVivo 11, a qualitative data analysis software program, was employed for the interview data analysis.

Table 1 summarizes the demographic information of the 18 interview participants. Participants were assigned codes P1–P18. Of the 18 interviewees, five were female and 13 were male. To conduct a comparison analysis, we took steps in the recruitment to ensure that interviewees were from two main academic areas: "hard sciences" comprising of sciences, computing and engineering, and "non-hard sciences" comprising of humanities, arts, business and social sciences. We also ensured that participants were from one of the two academic career levels (faculty members and non-faculty members).

Table 1. Demographics of the interview participants (N = 18).

	Total	%	Participants
Gender			
Female	5	28	P1, P11, P15, P16, P18
Male	13	72	P2, P3, P4, P5, P6, P7, P8, P9, P10, P12, P13, P14, P17
Discipline			
"Hard Sciences" (sciences, computing and engineering)	9	50	P1, P2, P3, P4, P5, P6, P7, P8, P9
"Non-hard Sciences" (humanities, arts, business and social sciences)	9	50	P10, P11, P12, P13, P14, P15, P16, P17, P18
Career level			
Faculty	6	33	P7, P8, P9, P16, P17, P18
Non-faculty	12	67	P1, P2, P3, P4, P5, P6, P10, P11, P12, P13, P14, P15

4   Results

In the following section, the interview results pertaining to the two research questions are presented.

4.1   RQ1: How Do NTU Researchers Promote Their Research Outputs?

Of the 18 participants interviewed, almost all of them (17) mentioned that they attend academic conferences in their fields, hold presentations and present posters to promote their research outputs. Publishing research findings in journals was regarded as another important venue to promote research by 16 respondents except P12 and P14. One faculty member (P16) stated that *"...present findings in conferences, publish in journals. They are effective in that they are likely to reach the intended audience. My strategy is choosing appropriate journals or conferences..."*. Besides, three participants (P11, P12 and P18) from non-hard sciences pointed out that networking with other researchers at workshops and research seminars was also an ideal venue to disseminate their research work.

Apart from traditional methods, it should be noted that some researchers used social media in their academic life. Half of the 18 scholars (P1, P2, P3, P4, P8, P10, P12, P15 and P17) said that they participate in academic social networks like ResearchGate and Google Scholar, and they believed that online platforms could be an alternative method for research dissemination, where researchers were able to create their own personal page and update their latest publications. One interviewee (P9) stated *"...at ResearchGate I get regularly updates like who is citing my paper or there is a download request for the full version and they also provide some statistics. Also, Google Scholar is quite nice, then I can directly see who cited you, where it was published, and things like that..."*. For most researchers, Facebook and Twitter are used mainly for private communication with friends and family members. However, one faculty staff (P8) from hard sciences preferred to use social media platforms such as Facebook and Twitter to promote research findings. The interviewee was planning to have a Twitter account for students to share tweets about publications to get more attention from the general public. Figure 1 describes the common methods used by the 18 NTU scholars to promote their research outputs.

4.2   RQ2: How Do NTU Researchers Evaluate Research Outputs?

The interview results provide insights to understanding which metrics are currently used by NTU scholars to evaluate research works. Based on the interview analysis, all 18 participants agreed with using traditional metrics to evaluate research outputs and they were quite familiar with some of them, especially citation count and the JIF. They believed that these metrics have already been tested by previous scholars and widely used over a long period, and were thus acceptable as a standard way to evaluate research outputs in academia. One interviewee (P1) stated: *"I think the most professional method for research evaluation is using traditional metrics because this is the only way other people will agree with your research works."* Further, to determine the quality of a research paper, most interviewees (16 out of 18) preferred to check in

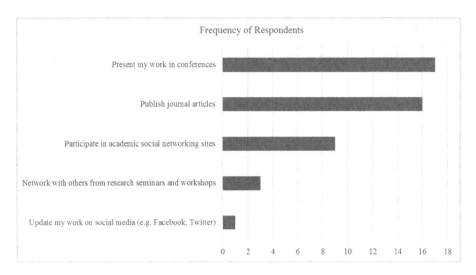

Fig. 1. Strategies to promote research outputs.

which journal the research paper was published, the given journal's impact factor and the ranking of the journal in the relevant discipline. One researcher (P15) stated: *"Regarding the quality of an article, I would firstly look at the impact factor of the journal, the university or the institute where the researcher is from and the researcher's previous publications or citations. Mostly based on the traditional metrics"*. Therefore, traditional metrics especially citation count and the JIF were highly accepted and used by NTU researchers, according to the responses of the participants.

However, more than half of the interviewees (10 out of 18) (P4, P6, P8, P9, P11, P12, P13, P14, P15 and P18) still argued that some limitations exist in the traditional ways of evaluating research outputs. For instance, research papers from certain academic fields such as computer science and engineering are likely to get a much higher citation count than those from social sciences. This might be a reason for researchers from certain minor research areas not to consider traditional metrics for research evaluation. One scholar (P18) from non-hard science said: *"Maybe in some fields like computer science or engineering, they can easily get a lot of citations, but for social science, it is very hard to get high citation counts and media to know your work. So, to some subjects, it is not a fair game..."*. In addition, two respondents (P7 and P9) were skeptical about the reliability of citation counts due to the possibility of gaming them. For example, one researcher (P7) stated: *"...I think that there are limitations such as it is possible to game the current statistics, just based on Google Scholar or WoS..."*. Therefore, it is not sufficient to take only traditional metrics into consideration when evaluating the quality of research findings as well as when evaluating the contribution of an academic researcher. Considering additional dimensions in the evaluation process would help to give a more holistic view of the impact of the research work.

As previously mentioned, some interviewees said that they had noticed the existence of altmetrics when they were using academic social media platforms Research-Gate and Google Scholar to promote their research. They might occasionally judge whether their works were popular by viewing the comments and downloads of their publications, but they had not used them formally as indicators for research evaluation. One third of the interviewees mentioned that altmetrics could be a good tool to complement traditional metrics. For example, one researcher (P8) from a hard science discipline stated *"I think we should go by both metrics, the traditional metrics as well as the altmetrics. There should be a weightage given to the altmetric score, especially in cases where the traditional metrics do not give enough information about the research outputs of a researcher. For example, in tenure and promotion, if people look at the traditional metrics and they feel this is not conclusive enough, then they should look at altmetrics and see if that gives them a more complete picture of the research impact of the researcher"*. Also, a few interviewees (P11, P15 and P17) from non-hard sciences indicated that they were willing to use altmetrics in the future if NTU were to establish relevant policies on the adoption of altmetrics in the research evaluation process. One respondent (P17) from social sciences stated *"Personally I don't care so much for now as that is not part of my performance evaluation. If NTU says you need a certain kind of media exposure to get your bonus, to get your tenure promotion, I think of course everyone will pay attention to it"*. Thus, this leads to the assumption, that probably most NTU researchers have not yet used social media metrics as criteria to assess the quality of research outputs.

Several reasons could explain why altmetrics have not yet been adopted by the NTU researchers interviewed. First, they were more familiar with traditional methods to decide whether their research findings had significant impact. For instance, one faculty staff (P17) from non-hard sciences indicated: *"I rely more on traditional metrics like citation counts and H-index. They have been here for so many years and they have specific criteria for filtering research outputs. The peer-review process is a good method to check the quality. Citation counts is also a good indicator"*. Second, participants were also concerned about the quality of altmetrics. For example, readers on social media may not spend as much time and effort on reading the paper and may give comments without careful consideration. One research staff (P14) argued: *"altmetrics such as the number of likes and retweets cannot mean so much. Maybe some people like it only because they want to show they have seen or noticed it"*. Researchers were somewhat skeptical that social media is not about quality, but more about popularity. On the one hand, they found that if more effort was put into promoting works online, then one would probably receive more likes or tweets. The number of likes only means the paper has received more public attention, but it is not necessarily related to the quality of the paper. On the other hand, they found that altmetrics seem only suitable for research outputs on certain topics of interest to the general public. For example, common people could be more interested in research findings related to their daily life rather than in advanced theories with intricate terminology. Hence, the research works related to public issues or social events may get a higher altmetrics from a general audience. In totality, the interviewees found that while traditional metrics are used more in academia, among colleagues and researchers who work in the same field, altmetrics could serve as a signal for non-experts to determine the popularity of a publication by showing quantitative comparisons on social media platforms.

5 Discussion and Conclusion

Understanding methods to promote research works and metrics to evaluate research outputs is of importance as this could help both researchers and research institutions to assess the significance of research outputs from a more comprehensive perspective. This study adopted a qualitative method to investigate how NTU scholars promote their research and what metrics are used to assess their research impact. Currently, attending academic conferences and publishing paper in journals are the two main methods to promote research findings among NTU researchers. With the popularity of social media in academia, NTU researchers have begun to use academic social networking sites such as ResearchGate and Google Scholar, as an alternative way to promote their research. However, currently, the usage of social media tools such as Facebook and Twitter, for promoting research is still low among participants since these tools are used more for communication with friends and families.

In terms of metrics used for research evaluation, the results of this study reveal that most participants use traditional metrics such as citation counts and JIF frequently in their professional lives. These metrics serve as good indicators to determine the quality of research outputs. NTU researchers have noticed the existence of altmetrics in various social media platforms; however, they have not considered using them in their research career. The strong reliance on traditional methods and quality concerns regarding altmetrics were two main hindrances for the usage of altmetrics by the interviewees. Altmetrics were however found to be a potential tool to complement traditional metrics as they could make the research evaluation process more comprehensive. We also found that a few researchers were willing to use social media and altmetrics given that they are supported by the university.

This case study was based on 18 semi-structured interviews with researchers from NTU, Singapore. These participants provided us with a wealth of knowledge about how they promote their research findings and how they evaluate research outputs. The results could help understanding the effective methods for research promotion and whether new metrics on social media platforms could be incorporated into the traditional research evaluation methods. The limitation of this study is the limited sample size due to the resource and time constraints. Future studies should include more researchers from different fields and different institutions to give a more comprehensive picture about how metrics are used for research evaluation in academia.

Acknowledgements. This research is supported by the National Research Foundation, Prime Minister's Office, Singapore under its Science of Research, Innovation and Enterprise programme (SRIE Award No. NRF2014-NRF-SRIE001-019). Also, we would like to acknowledge and thank all our interview participants, who kindly volunteered their time and professional opinions to our study.

Appendix

Interview Questions

1. How do you usually promote your research? What types of strategies do you use?
   a. Why do you choose such ways to promote your research?
   b. Do you feel they are effective?
2. What do you think of disseminating research outputs on social media? Why?
3. How do you keep track of your research outputs?
4. What do you think of the way research outputs are being evaluated among your colleagues?
5. Until now, how do you evaluate your research outputs? How often do you evaluate them?
6. Why do you evaluate your research outputs?
7. If you had a magic wand, what would be your ideal way of evaluating research outputs?
8. Can you tell me about any problems that you encounter in evaluating research outputs?
   a. How did you learn about these problems?
   b. Why is this considered a problem?
   c. How did you solve it?
9. How do you determine the quality of an article/researcher/journal?
10. What kind of impact do you think traditional metrics/altmetrics have in evaluating research outputs?
11. What do you think are some of the characteristics of traditional metrics/altmetrics that make them useful in evaluating research outputs?
12. How do you think traditional metrics/altmetrics will help in evaluating research outputs?
13. Are you hesitant about using any traditional metrics/altmetrics to evaluate research outputs? Why?
14. How do you think social media has changed the way research outputs are evaluated?
15. How do you think the emergence of alternative metrics has changed your research career?
16. Do you have any other comments? Is there anything else you'd like to tell me?

   Clarifying questions:

- Can you give me some examples?
- Can you elaborate more on this?
- Can you tell me anything else?

References

1. Ovadia, S.: When social media meets scholarly publishing. Behav. Soc. Sci. Libr. **32**(3), 194–198 (2013)
2. Schnitzler, K., Davies, N., Ross, F., Harris, R.: Using Twitter™ to drive research impact: a discussion of strategies, opportunities and challenges. Int. J. Nurs. Stud. **59**, 15–26 (2016)
3. Gruzd, A., Staves, K., Wilk, A.: Connected scholars: Examining the role of social media in research practices of faculty using the UTAUT model. Comput. Hum. Behav. **28**(6), 2340–2350 (2012)
4. Harley, D., Acord, S.K., Earl-Novell, S., Lawrence, S., King, C.J.: Assessing the future landscape of scholarly communication: an exploration of faculty values and needs in seven disciplines. Final Report. Center for Studies in Higher Education. University of California, Berkley, CA (2010)
5. Weller, K.: Social media and altmetrics: an overview of current alternative approaches to measuring scholarly impact. In: Incentives and Performance, pp. 261–276 (2015)
6. Alotaibi, N.M., Guha, D., Fallah, A., Aldakkan, A., Nassiri, F., Badhiwala, J.H., et al.: Social media metrics and bibliometric profiles of neurosurgical departments and journals: Is there a relationship? World Neurosurg. **90**, 574–579 (2016)
7. Kratz, J.E., Strasser, C.: Researcher perspectives on publication and peer review of data. PLoS ONE **10**(2), e0117619 (2015)
8. Bordons, M., Fernández, M., Gómez, I.: Advantages and limitations in the use of impact factor measures for the assessment of research performance. Scientometrics **53**(2), 195–206 (2002)
9. Haustein, S., Siebenlist, T.: Applying social bookmarking data to evaluate journal usage. J. Informetr. **5**(3), 446–457 (2011)
10. Sud, P., Thelwall, M.: Evaluating altmetrics. Scientometrics **98**(2), 1131–1143 (2014)
11. Erdt, M., Nagarajan, A., Sin, S.C.J., Theng, Y.L.: Altmetrics: an analysis of the state-of-the-art in measuring research impact on social media. Scientometrics **109**(2), 1117–1166 (2016)
12. Loach, T.V., Evans, T.S.: Ranking journals using altmetrics. arXiv preprint arXiv:1507. 00451 (2015)
13. Sugimoto, C.R., Work, S., Larivière, V., Haustein, S.: Scholarly use of social media and altmetrics: a review of the literature. J. Assoc. Inf. Sci. Technol. **68**(9), 2037–2062 (2017)
14. Bonetta, L.: Scientists enter the blogosphere. Cell **129**(3), 443–445 (2007)
15. Nicholas, D., Rowlands, I.: Social media use in the research workflow. Inf. Serv. Use **31**(1–2), 61–83 (2011)
16. Stutzman, F., Capra, R., Thompson, J.: Factors mediating disclosure in social network sites. Comput. Hum. Behav. **27**(1), 590–598 (2011)
17. Thelwall, M.: A brief history of altmetrics. Res. Trends **37**, 3–4 (2014)
18. Gruzd, A., Staves, K., Wilk, A.: Tenure and promotion in the age of online social media. Proc. Am. Soc. Inf. Sci. Technol. **48**(1), 1–9 (2011)

Scientific vs. Public Attention: A Comparison of Top Cited Papers in WoS and Top Papers by Altmetric Score

Sumit Kumar Banshal[1(✉)], Aparna Basu[1], Vivek Kumar Singh[2], and Pranab K. Muhuri[1]

[1] Department of Computer Science, South Asian University,
New Delhi 110021, India
sumitbanshal06@gmail.com
[2] Department of Computer Science, Banaras Hindu University,
Varanasi 221005, India

Abstract. Alternative metrics or altmetrics are article level metrics that have been used to quantify the attention given to scholarly papers in online fora or social media. In a way it was sought to replace journal level metrics such as the Impact Factor and also to see if altmetric scores could predict highly cited papers. Early studies, done soon after altmetrics was proposed in 2010, were somewhat premature as the use of social media was not widely prevalent, and their results may no longer be relevant at the present time when use levels have risen significantly. In 2016, Altmetric.com has tracked over 17 million mentions of 2.7 million different research outputs. Of these, the top 100 most-discussed journal articles of 2016 were presented with details of journals in which the research was published, affiliations of authors, fields, countries, etc. Our attempt is to obtain a bibliometric profile of these papers with high altmetric scores to see the underlying patterns and characteristics that motivate high public attention. In parallel we have analyzed top 100 highly cited papers from the same year. Our objective in this empirical study is to examine the similarities and distinguishing features of scientific attention as measured by citations and public attention in online fora.

A significant finding is that there is very little overlap between very highly cited papers and those that received the highest altmetric scores.

Keywords: Altmetrics · Social media · Impact analysis · Web of science
Science Citation Index-Expanded

1 Introduction

Altmetrics or 'alternative metrics' is an attempt to capture the degree to which certain scholarly items such as articles, book chapters, etc. get the attention of their readers [1]. The metric can be obtained in terms of number of reads, number of downloads etc. and has been conceived as an indicator of 'quality'. In an earlier time, quality of an article was measured by number of citations, and also in terms of the 'impact factor' or the quality of the journal in which it was published. However, 'journal impact factor' was

© Springer Nature Singapore Pte Ltd. 2018
M. Erdt et al. (Eds.): AROSIM 2018, CCIS 856, pp. 81–95, 2018.
https://doi.org/10.1007/978-981-13-1053-9_7

deemed unsatisfactory as a metric for individual papers, while the accumulation of citations was a slow process, and it could take several years to identify highly cited papers or authors. This put younger scholars at a disadvantage, and the shortcomings of citation based evaluation were more keenly felt in evaluation of scholars for recruitment or promotion to higher levels, say, in university or other institutions of higher education and research. Funders were also looking at the prospects of using altmetrics. The measures could help young researchers who have fewer citations than their more senior colleagues[1].

The emergence of online 'social media' like Twitter, blogs, Academic social networks such as Mendeley, Research Gate etc. gave the possibility of recording readers' responses to what they read. Altmetrics.com[2], a Data Science company, combines responses in a variety of media to obtain an 'Altmetric Attention Score'[3] (see Table 1). This score is given by many journals, e.g. Nature, PLOS, etc. alongside their papers. Social media has a multiplicative effect in that people tend to read articles that have been mentioned by others – an example of 'cumulative advantage' [2] See dissertation. Altmetrics intend to measure scholarly impact that is not necessarily captured by traditional, citation based metrics. These measures are mostly derived from the web and social media. Altmetrics have already started to become a part of today's scholarly communication and a measure of its impact. Scientific journals are increasingly providing the number of tweets, Facebook likes and Mendeley readers as well as other social media mentions of their articles. At the same time, researchers have begun to present altmetrics in their CVs and universities and funders are starting to consider the use of social media metrics to better understand the impact of their scientific output

In 2016, Altmetric has tracked over 17 million mentions of 2.7 million different research outputs. Of these, the top 100 most-discussed journal articles of 2016 were presented with details of journals in which the research was published, affiliations of authors, fields, countries, etc.[4] Our attempt in this paper is to obtain a bibliometric overview of this selection. We also compare our results with the top 100 highly cited papers from the same year (2016) from Web of Science. The objective is to note the differences (and possible similarities) between the citation based selection and the altmetrics based one. We end with a discussion on the promise of altmetrics.

2 Objectives

Our objective is to do a bibliometric analysis of 100 papers with the highest Altmetrics scores in 2016 (published by Altmetric.com) to see what are the characteristics of papers that obtain a large number of online mentions. We examine the titles of the papers, the corresponding subject categories, the journals in which they are published

---

[1] https://www.nature.com/news/funders-drawn-to-alternative-metrics-1.16524.

[2] https://www.altmetric.com/.

[3] https://help.altmetric.com/support/solutions/articles/6000060969-how-is-the-altmetric-attention-score-calculated.

[4] https://www.altmetric.com/top100/2016/.

and the countries from which most of these papers are published. We also do a parallel analysis of the 100 most cited papers in Web of Science along the same lines. This exploratory analysis is undertaken to see if some general underlying patterns can be found between the citation processes.

3   Altmetrics

Social network usage (e.g. ResearchGate, Academia, Twitter, Facebook etc.) saw an exponential growth in the last decade. Since the time of introduction, social networks attracted a lot of researchers. Scientific researchers are not only becoming familiar with these platforms but also using them to share their research work. Therefore a huge volume of data on scholarly articles is generated through social networks. Both altmetrics and citation indexing thus rely on collective intelligence, or the wisdom of the crowds, to identify the most relevant scholarly works [3]. This huge volume data can give a different, new and instant view of the importance of articles. Hence, modern technologies or approaches are needed to assess this new form of impact. This assessment technique is commonly termed or known as alternative metrics or 'Altmetrics'. Altmetrics is the study and use of scholarly impact measures based on activity in online tools and environments. Altmetrics is not only the field of study of scholarly data but also metrics itself, which makes it different from "bibliometrics" or "scientometrics" [4]. This introduced social network data, also introduced new tools and new possibilities of measuring of scholarly articles [5, 6]. The newly termed metrics based on online social network usage was first proposed in 2010 and termed as "altmetrics" [1]. Specifically, Altmetrics is the evolution from old-style bibliometrics in the direction of a new field and metrics in the 21st century [8].

As social networks are becoming universal [9], researchers are gradually moving into altmetrics research [10]. This topic even seems to have overtaken the h-index [11], a research topic which was one of the most prominent as well as most influential research topic in recent past [12]. Altmetrics are constructed from events on social networks [13], which are generated from actions made on scholarly articles. There are numerous groupings for events [14]. In 'article-level metrics' (ALMs, [15]), views, downloads, clicks, notes, saves, tweets, shares, likes, recommends, tags, posts, trackbacks, discussions, bookmarks, and comments are counted, rather than just citations of a paper in a database such as Scopus (Elsevier), or by a publisher such as the Public Library of Science (PLOS, [16]) [17]. There are several methods by which altmetrics events can be derived from a dataset [18]. The growth of the idea of altmetrics has been accompanied by evolution in the variety of web-based tools intended at apprehending and following a widespread range of a researcher's productivity by collecting altmetrics data through a comprehensive range of sources [19, 20]. The widespread usage of the social web by scholars has also led to studies of altmetrics and its relation to well-known impact metrics such as citation analysis [21]. Most of these studies have found some degree of correlation between altmetrics and citation scores, signifying that these two methods are somehow correlated but not the same. There is some work done on different social network events as well, where researchers try to establish a relation or try to develop a wide range of qualitative & quantitative aspects [22]. There are several research papers where different events, different datasets [23] are discussed. In this research area, there are so many pages yet unturned though. Here, one approach is presented to visualize the agreements or disagreement (if any) between the well-known, widely-used impact i.e., citations and the newly designed attention metrics.

4 Data and Analysis

We have selected our data from ALTMETRICS 2016 report, which lists 100 scientific papers with the highest Altmetric Attention Scores[5]. The Score is computed from mentions in social media, such as Twitter, News, Facebook etc. (see Table 1 for the full list).

Table 1. Sources for altmetric attention score

Source	Items	%
Twitter	101,403	75.4
News	23,104	17.2
Facebook	5,020	3.7
Blogs	2,273	1.7
Google+	2,052	1.5
Reddit	268	0.2
Wikipedia	190	0.1
Video	69	0.1
Peer Review	42	0.0
F1000	36	0.0
Q&A	20	0.0
Policy	16	0.0
Total	134493	100

The largest part of the score in ALTMETRICS 2016 comes from Twitter data (75%) and the rest from other sources. In the calculation process of the Altmetric Attention Score, the weighted count methodology is followed. The weights are given according to their possibilities of getting public or researcher's attention[6].

We have looked at the country wise distribution of the Altmetric Attention Score, the journal distribution and the affiliation to understand what kinds of papers receive the most attention. For comparison, we have selected the 100 most cited papers published in 2016, as indexed in Web of Science, and analysed it in a similar way.

4.1 Papers

Let us first look at the papers common to both lists, i.e., they are both highly cited and have also received high altmetric scores (Table 2). Surprisingly, there were only 12

---

[5] https://www.altmetric.com/top100/2016/#about=show.

[6] https://help.altmetric.com/support/solutions/articles/6000060969-how-is-the-altmetric-attention-score-calculated.

papers common to both lists that received relatively high citations and altmetric scores (Table 2). Seven out of 12 papers fell in the category of Medical and Health Sciences. They obtained 62.7%, of the total Altmetric Attention Score (18,864 out of a total of 30,090 for 12 papers) and a total citation of 57.1% (3382 citations out of a total of 5984 citations for 12 papers).

Table 2. Papers present in both the Thomson Reuter's top 100 cited papers in web of science and the top 100 papers in ALTMETRICS 2016

No.	Title	Journal	Alt_score[a]	Citation[b]	Ratio alt-score/citation	Subject
1	Observation of Gravitational Waves from a Binary Black Hole Merger	Physical Review Letters	4660	1327	3.51	Physical Sciences
2	The Third International Consensus Definitions for Sepsis and Septic Shock (Sepsis-3)	JAMA-Journal Of The American Medical Association	2471	971	2.54	Medical & Health Sciences
3	Emergence of plasmid-mediated colistin resistance mechanism MCR-1 in animals and human beings in China: a microbiological and molecular biological study	Lancet Infectious Diseases	1708	603	2.83	Biological Science
4	Zika Virus Associated with Microcephaly	New England Journal Of Medicine	2464	572	4.31	Medical & Health Sciences
5	Guillain-Barre Syndrome outbreak associated with Zika virus infection in French Polynesia: a case-control study	Lancet	2171	441	4.92	Medical & Health Sciences
6	Zika Virus and Birth Defects - Reviewing the Evidence for Causality	New England Journal Of Medicine	3753	406	9.24	Medical & Health Sciences
7	Mastering the game of Go with deep neural networks and tree search	Nature	3047	343	8.88	Information & Computer Sciences
8	Cluster failure: Why fMRI inferences for spatial extent have inflated false-positive rates	Proceedings Of The National Academy Of Sciences Of The United States Of America	1847	289	6.39	Medical & Health Sciences
9	The ASA's Statement on p-Values: Context, Process, and Purpose	American Statistician	1811	269	6.73	Research & Reproducibility

(continued)

Table 2. (*continued*)

No.	Title	Journal	Alt_score[a]	Citation[b]	Ratio alt-score/citation	Subject
10	Zika Virus in the Americas - Yet Another Arbovirus Threat	New England Journal Of Medicine	2530	260	9.73	Medical & Health Sciences
11	Zika Virus Infection in Pregnant Women in Rio de Janeiro	New England Journal Of Medicine	1672	220	7.60	Medical & Health Sciences
12	Zika Virus Infects Human Cortical Neural Progenitors and Attenuates Their Growth	Cell Stem Cell	1956	223	8.77	Medical & Health Sciences

[a]From ALTMETRICS 2016; [b]Web of Science 2016

Half of the common papers, or six out of eight Medical Science papers, were about the Zika virus outbreak and its various aspects, which obviously got significant scientific and public attention in 2016. The other Medical Science papers receiving both scientific and public attention were on guidelines for identification of Sepsis and on errors in inferences from fMRI data.

The papers common to both lists had only singular papers in other categories. In Biological Sciences, there was a paper from China on emergence of plasmid-mediated colistin resistance in China. In the Physical Sciences category, the paper on Gravitational waves received both high citations and a high alt-score. In Information and Computer Sciences, the game of GO received a lot of attention. In Research and Reproducibility, the American Statistical Association's statement on p-values received both scientific and public attention.

Table 3. Papers present in both the Thomson Reuter's top 100 cited papers in Web of Science and the top 100 papers in ALTMETRICS 2016

Position	Title	Journal	Score	Open access	Content type	Subject
1	United States Health Care Reform: Progress to Date and Next Steps	JAMA	8063	Free	Special Communication	Studies in Human Society
2	Medical error—the third leading cause of death in the US	British Medical Journal	4912	No	Analysis	Studies in Human Society
3	Observation of Gravitational Waves from a Binary Black Hole Merger	Physical Review Letters	4660	Yes	Article	Physical Sciences
4	Evidence for a Distant Giant Planet in the Solar System	The Astronomical Journal	4319	No	Article	Physical Sciences
5	Sugar Industry and Coronary Heart Disease Research: A Historical Analysis of Internal Industry Documents	JAMA Internal Medicine	4297	No	Special Communication	Research & Reproducibility

(*continued*)

Table 3. (*continued*)

Position	Title	Journal	Score	Open access	Content type	Subject
6	Zika Virus and Birth Defects — Reviewing the Evidence for Causality	New England Journal of Medicine	3753	Yes	Special Report	Medical & Health Sciences
7	The Association Between Income and Life Expectancy in the United States, 2001–2014	JAMA	3735	No	Special Communication	Studies in Human Society
8	Effect of Wearable Technology Combined With a Lifestyle Intervention on Long-term Weight Loss: The IDEA Randomized Clinical Trial	JAMA	3101	No	Original Investigation	Medical & Health Sciences
9	Mastering the game of Go with deep neural networks and tree search	Nature	3047	No	Article	Information & Computer Sciences
10	The new world atlas of artificial night sky brightness	Science Advances	3020	Yes	Article	Physical Sciences

Table 4. Papers present in both the Thomson Reuter's top 100 cited papers in Web of Science and the top 100 papers in ALTMETRICS 2016

No	Title	Journal	Total citations
1	Cancer Statistics, 2016	Ca-A Cancer Journal For Clinicians	3810
2	Cancer Statistics in China, 2015	Ca-A Cancer Journal For Clinicians	1259
3	MEGA7: Molecular Evolutionary Genetics Analysis Version 7.0 for Bigger Datasets	Molecular Biology And Evolution	1331
4	Observation of Gravitational Waves from a Binary Black Hole Merger	Physical Review Letters	1323
5	Heart Disease and Stroke Statistics-2016 Update A Report From the American Heart Association	Circulation	1039
6	The Third International Consensus Definitions for Sepsis and Septic Shock (Sepsis-3)	JAMA-Journal Of The American Medical Association	880
7	Planck 2015 results XIII. Cosmological parameters	Astronomy & Astrophysics	872
8	The Cambridge Structural Database	Acta Crystallographica Section B-Structural Science Crystal Engineering And Materials	755

(*continued*)

Table 4. (*continued*)

No	Title	Journal	Total citations
9	2015 ESC Guidelines for the management of acute coronary syndromes in patients presenting without persistent ST-segment elevation Task Force for the Management of Acute Coronary Syndromes in Patients Presenting without Persistent ST-Segment Elevation of the European Society of Cardiology (ESC)	European Heart Journal	711
10	Analysis of protein-coding genetic variation in 60,706 humans	Nature	675

The list in Table 4 of the most highly cited papers in 2016 shows that they are of a more technical nature than the top altmetrics papers. The two most highly cited papers are from the Cancer Statistics in China in two successive years. Another is on MEGA7, a software in genomic analysis, a paper on Heart Disease and Heart stroke statistics from USA, guidelines for the management of acute coronary syndromes, analysis of protein coding genetic variation in ~60,000 humans, on the Cambridge Structural Database, and International definitions of sepsis and septic shock. Apart from this, there are two physics papers on Gravitational waves and on Planck 2015: Cosmological parameters. Clearly these are of a highly technical nature and would be of interest to a professional audience but not to society at large.

Let us focus our attention now on different aspects of these high visibility papers, a total of 200 with 12 common papers. We can ask which subject areas appeared to be discussed more or get the most mentions, which journals are most seen and commented upon?, which countries or institutions do these highly visible papers come from? In the following sections we address some of these questions.

4.2 Subjects

The Medical and Health Sciences received the maximum attention with 49 papers in out of 100 in the ALTMETRICS 2016 report. This was followed by the Biological Sciences (14) and Studies in Human Society (12). This shows that public attention is captured by problems that they can relate to or that is likely to affect them, as shown in Table 5. This point was made several years ago when a very large survey on public understanding of science was done amongst people of average literacy in India [23]. When asked questions on scientific matters, they could relate much better to themes which directly affected them or were close-by than to themes that were remote or abstract.

A comparison of subjects in ALTMETRICS 100 and top 100 cited papers (Fig. 1) shows that Medical Health, Biology, History and Archaeology, Earth and Environmental Science received more public attention, whereas Physical Sciences, Materials

Sciences and Study of Human Societies appear to have gotten more scientific attention i.e., more papers in top 100 cited list and fewer papers in the Altmetrics list. (The subject areas in the two lists had to be harmonized, by combining WoS categories. This may have given rise to some errors, for example Studies in Human Societies had no good correlations among the WoS categories, and was obtained by combining Energy Fuels and Science Technology and other Studies.)

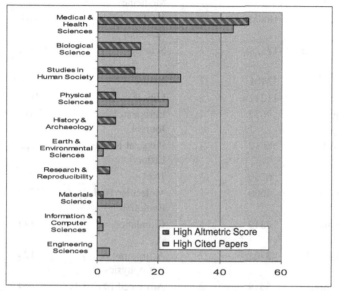

*Authors' elaboration of data from ALTMETRIC 2016 and Web of Science 2016.

Fig. 1. Relative distribution of papers in different subject areas indicating more public attention (more papers with high altmetric score) or more scientific attention (more papers with high citation)

Table 5. Subject areas that received the most public/scientific attention in 2016

Subjects	High altmetric score	High cited papers
Medical & Health Sciences	49	44
Biological Science	14	11
Studies in Human Society	12	27
Earth & Environmental Sciences	6	2
History & Archaeology	6	
Physical Sciences	6	23
Research & Reproducibility	4	
Materials Science	2	8
Information & Computer Sciences	1	2
Engineering Sciences	0	4
Total	100	121

Source: Author's elaboration of data from ALTMETRIC 2016; Web of Science 2016

Table 6. Top 10 journals from the Altmetrics 2016 list and from top cited papers-WoS 2016

Journal (Altmetrics.com)	Alt_Score	Paper	Journal (Web of Science)	Citation	Papers
Nature	39391	18	Ca-A Cancer Jnl For Clinicians	6021	3
JAMA	22921	7	New England Jnl Of Medicine	3406	10
Science	18806	9	Nucleic Acids Research	2816	7
New England Jnl of Medicine	17528	7	Science	2379	8
PNAS	13401	7	Nature	2261	7
The Lancet	11311	5	Lancet	2153	6
JAMA Internal Medicine	9672	4	European Heart Journal	1816	5
British Medical Journal	8612	3	Physical Review Letters	1722	2
Current Biology	6226	3	JAMA-	1700	4
Physical Review Letters	4660	1	Molecular Biology & Evolution	1379	1
The Astronomical Journal	4319	1	Circulation	1377	2
Annals Of Internal Medicine	3765	2	Astronomy & Astrophysics	1124	2
Scientific Reports	3579	2	Advanced Materials	864	3
JAMA Pediatrics	3558	2	Blood	860	2
Science Advances	3020	1	Lancet Infectious Diseases	840	2

Sorted by Citations and Altmetric Score

4.3 Journals

A bibliometric analysis of the journal data should show which journals carry the most highly cited papers and which publish articles that get the highest altmetric scores.

The first thing we noted is that the journals carrying the top 100 cited papers and those carrying papers with the highest altmetrics scores are also quite different. There is an overlap of a mere 10 journals, or 10%, common to the two lists (Table 7).

We notice that the Altmetrics list is more concentrated at the upper end with 18 papers in Nature. Compared to that, the cited papers are more evenly dispersed. Physical review Letters had the highest Altmetric Attention Score per paper.

We note from Table 8, that the journals for the two data sets is largely disjoint, and journals that publish papers with high citations are not the same as those that get high altmetric scores. The highest number of citations per paper was obtained by Cancer Journal for Clinicians followed by Molecular Biology and Evolution and Physical Review Letters.

Table 7. Journals publishing papers common to top altmetrics and top cited lists

Journal	Alt-Papers	Alt_Score	Avg score per paper	WoS-papers	Citation	Avg citation per paper
Nature	18	39391	2188	7	2261	323
Science	9	18806	2090	8	2379	297
JAMA	7	22921	3274	4	1700	425
Physical Review Letters	1	4660	4660	2	1722	861
Science Advances	1	3020	3020	1	434	434
Mmwr Recommendations and Reports	1	1967	1967	1	228	228
Cell Stem Cell	1	1956	1956	1	223	223
European Heart Journal	1	1785	1785	5	1816	363
Lancet Infectious Diseases	1	1708	1708	2	840	420
Cell Metabolism	1	1682	1682	1	223	223
Total	41	97896	2387.7	32	11826	370

Table 8. Journals in top cited list with highest citations per paper

Journals	Total Cites	Papers	% Cites	Avg Cites per Paper
Ca-A Cancer Journal For Clinicians	6021	3	14.9	2007[a]
Molecular Biology And Evolution	1379	1	3.4	1379[a]
Physical Review Letters	1722	2	4.3	861
Acta Crystallographica Section B-Structural Science Crystal Engineering And Materials	755	1	1.9	755[a]
Circulation	1377	2	3.4	689[a]
Astronomy & Astrophysics	1124	2	2.8	562[a]
Acta Neuropathologica	559	1	1.4	559[a]
Science Advances	434	1	1.1	434
Blood	860	2	2.1	430[a]
JAMA-Journal Of The Amer Med Assoc	1700	4	4.2	425
Lancet Infectious Diseases	840	2	2.1	420
Chest	417	1	1.0	417[a]
Others	....	....	....	< 403
Total	40304	100	100	403.0

[a]Journals not common to both Altmetric and Cited Lists

4.4 Countries

In this section, we try to identify countries where authors of these high cited or high altmetric score papers reside. (Tables 9 and 10)

From Table 9 we see that a large percentage among the top 100 cited papers are authored in the USA. England and Germany author 29% and 26% of the papers respectively. France and Canada author 22% of the papers each. Other countries author less than 20% in the top 100 elite cited set. The distribution across countries for high alt-score papers is similar but less highly concentrated in USA and UK (Table 10).

Table 9. Distribution of citations across countries based on author affiliation

Countries/Territories	TP	% of 100	Total citations	Citations per papers
USA	64	64	29730	437.206
England	29	29	11936	442.074
Germany	26	26	10109	404.36
France	22	22	10924	520.19
Canada	22	22	8530	406.19
Italy	19	19	7144	396.889
Australia	18	18	8334	490.235
Peoples R China	17	17	7728	483
Spain	16	16	6603	440.2
Switzerland	15	15	5468	420.615

Source: Author's elaboration of Web of Science 2016 Top 100 cited papers

Table 10. Distribution of author affiliations in ALTMETRICS 2016 across different countries

Country	TP	% of 100	Score	% Score	Alt_Score/Paper
United States	75	32.2	174686	32.4	2329
United Kingdom	33	14.2	69721	12.9	2113
Germany	14	6.0	31816	5.9	2273
Australia	12	5.2	26955	5.0	2246
France	8	3.4	18967	3.5	2371
Canada	6	2.6	14627	2.7	2438
Italy	5	2.1	13479	2.5	2696
Belgium	5	2.1	13035	2.4	2607
Netherlands	5	2.1	12769	2.4	2554
Switzerland	6	2.6	12415	2.3	2069

5   Results

A quick review of the titles in Table 3 indicates that these papers are of great significance not only to scientists or researchers but to a very wide class of people. The first paper with the highest mentions (score $\sim$ 8000) - United States Health Care Reform: Progress to Date and Next Steps, is by President Obama and is on a topic likely to be of interest to a very large set of people. The second paper is on medical error being a leading cause of death in the USA. These are categorized as studies in human society. Another paper in the same category relates to income and life-expectancy in the United States. There are three Physical Science papers on observation of gravitational waves, a world atlas of artificial night brightness, and evidence for a distant giant planet. The other papers in Medical Science are on the Zika virus, which garnered a lot of attention after the first outbreak in Brazil, and a clinical trial on the use of wearable technology for weight-loss. A paper in Computer Science was about the Chinese game GO, where a leading player was defeated by a computer. A paper in the Research and Reproducibility category related to coronary heart disease and the sugar industry.

Qualitatively speaking, this list of top 10 papers confirms that article level metrics definitely capture attention and immediate concern of many, and it may be inferred that these are works of real significance to people. Among the journals which were common to both the Altmetrics list and the WoS highly cited list in 2016, SCIENCE was the most highly cited journal with eight papers (5.14% of 43974 total citations) followed by NATURE with seven papers (5.14%). The corresponding figures for the Almetrics list are 18 papers from NATURE (17.4% of total score), seven papers from JAMA (10.2%), nine papers from SCIENCE (8.3%), seven papers from New England Journal of Medicine (7.8%), seven papers from PNAS (5.93%) andfive5 papers from the Lancet (5.0%). The rest contribute less than 5% to the top 100 list.

Interestingly, the highest citations per paper go to journals other than those common to both lists. In other words, the journals that draw public attention are not the ones that are highly cited except for a small number that receive attention from both the public and academics. In Table 6, we have 12 journals with average citations per paper, higher than the group average (403 cites per paper). Of these, eight journals do not receive enough public attention to be in the Altmetrics list but do get a lot of scientific attention in terms of high citations per paper. CA-A CANCER JOURNAL FOR CLINICIANS got over 2000 citations per paper, while MOLECULAR BIOLOGY AND EVOLUTION got over 1000 cites per paper. The remaining 4 journals that receive high scientific attention and also public attention are Physical Review Letters, Science Advances, JAMA and Lancet Infectious Diseases.

We note that papers authored in USA, UK and Germany were the most highly cited and also had the highest Altmetric Attention Score. China was the only Asian country that featured in either list. China was 8th in the list of highly cited countries.

6 Conclusion

We conclude from our exploratory analysis that the top 100 Altmetric 2016 papers form an almost disjoint set from the top 100 cited papers with only a small overlap of 12 papers. In addition, the journals that publish these papers are also nearly disjoint with an intersection of only 10 journals. The high cited journals appear to have more technical content while the high altmetric papers deal with issues that have more widespread concern and appeal. It does not appear that one can predict the other, as the processes involved appear to be disjoint. Western countries such as the USA, UK, and Germany dominate both altmetrics and citations. China is the only Asian country in the top, it is 8th on the list of citations. The citation process takes time during which a paper is read and only thereafter it may be incorporated as a reference in a paper. In altmetrics the response is immediate, and the papers are less technical or with greater mass appeal. One can think of the altmetric score as supplying additional information to the citation score.

References

1. Priem, J., Taraborelli, D., Groth, P., Neylon, C.: Altmetrics: A manifesto (2010). http://altmetrics.org/manifesto
2. Kim, H.S.: Attractability and virality: The role of message features and social influence in health news diffusion. University of Pennsylvania (2014)
3. Haustein, S.: Grand challenges in altmetrics: heterogeneity, data quality and dependencies. Scientometrics 108(1), 413–423 (2016)
4. Priem, J.: Altmetrics. Beyond bibliometrics: Harnessing multidimensional indicators of scholarly impact, journal? 263–88 (2014)
5. Wouters P., Costas, R.: Users, Narcissism and Control – Tracking the Impact of Scholarly Publications in the 21st Century. Image Rochester, New York (2012)
6. Adie, E., Roe, W.: Altmetric: enriching scholarly content with article-level discussion and metrics. Learn. Publish. 26(1), 11–17 (2013)
7. Roemer, R.C., Borchardt, R.: Introduction to Altmetrics. Libr. Tehnol. Reports 51(5), 5–10 (2015)
8. Bar-Ilan, J., Haustein, S., Peters, I., Priem, J., Shema, H., Terliesner, J.: Beyond citations: Scholars' visibility on the social web. In: Proceedings of the 17th International Conference on Science and Technology Indicators, Montreal, Quebec, Canada (2012). http://arxiv.org/abs/1205.5611/
9. Sud, P., Thelwall, M.: Evaluating altmetrics. Scientometrics 98(2), 1131–1143 (2014)
10. Hirsch, J.E.: An index to quantify an individual's scientific research output. Proc. Natl. Acad. Sci. U.S.A. 102(46), 16569–16572 (2005)
11. Piwowar, H., Priem, J.: The power of altmetrics on a CV. Bull. Assoc. Inf. Sci. Technol. 39, 10–13 (2013)
12. Zahedi, Z., Costas, R., Wouters, P.: How well developed are Altmetrics? Cross disciplinary analysis of the presence of "alternative metrics" in scientific publications (RIP). In: Proceedings of the 14th International Society of Scientometrics and Informetrics Conference, Vienna, Austria, pp. 876–884. Facultas Verlags und Buchhandels, Wien (2013)
13. Bornmann, L.: Do altmetrics point to the broader impact of research? An overview of benefits and disadvantages of altmetrics. J. Informetrics 8(4), 895–903 (2014)
14. Fenner, M.: What can article-level metrics do for you? PLoS Biol. 11(10) (2013)
15. Costas, R., Zahedi, Z., Wouters, P.: Do "altmetrics" correlate with citations? Extensive comparison of altmetric indicators with citations from a multidisciplinary perspective. J. Assoc. Inf. Sci. Technol. 66(10), 2003–2019 (2015)
16. Liu, J., Adie, E.: Five challenges in altmetrics: A toolmaker's perspective. Bull. Am. Soc. Inf. Sci. Technol. 39(4), 31–34 (2013)
17. Thelwall, M., Haustein, S., Larivière, V., Sugimoto, C.R.: Do altmetrics work? Twitter and ten other social web services. PLoS ONE 8(5) (2013)
18. Thelwall, M., Kousha, K.: ResearchGate: Disseminating, communicating, and measuring Scholarship? J. Assoc. Inf. Sci. Technol. 66(5), 876–889 (2015)
19. Haustein, S., Peters, I., Sugimoto, C.R., Thelwall, M., Larivière, V.: Tweeting biomedicine: an analysis of tweets and citations in the biomedical literature. J. Am. Soc. Inf. Sci. Technol. 65(4), 656–669 (2014)
20. Shema, H., Bar-Ilan, J., Thelwall, M.: Do blog citations correlate with a higher number of future citations? research blogs as a potential source for alternative metrics. J. Assoc. Inf. Sci. Technol. 65(5), 1018–1027 (2014)

21. Haustein, S., Thelwall, M., Larivière, V., Sugimoto, C.R.: On the relation between altmetrics and citations in medicine (RIP). In: Hinze, S., Lottmann, A. (eds.) Proceedings of the 18th International Conference on Science and Technology Indicators (STI), Berlin, Germany, pp. 164–166, 4–6 September 2013
22. Waltman, L., Costas, R.: F1000 recommendations as a potential new data source for research evaluation: a comparison with citations. J. Assoc. Inf. Sci. Technol. **65**(3), 433–445 (2014)
23. Raza, G., Singh, S., Dutt, B.: Public, science, and cultural distance. Sci. Commun. **23**(3), 293–309 (2002)

Field-Weighting Readership: How Does It Compare to Field-Weighting Citations?

Sarah Huggett[1(✉)] ⓘ, Chris James[2] ⓘ, and Eleonora Palmaro[2] ⓘ

[1] Elsevier, 3 Killiney Road, #08-01, Winsland House 1,
Singapore 239519, Singapore
s.huggett@elsevier.com
[2] Elsevier, Radarweg 29, 1043 NX Amsterdam, Netherlands

Abstract. Recent advances in computational power and the advancement of the internet mean that we now have access to a wider array of data than ever before. If used appropriately, and in conjunction with peer evaluation and careful interpretation, metrics can inform and enhance research assessment through the benefits of being impartial, comparable, and scalable. There have been several calls for a "basket of metrics" to be incorporated into research evaluation. However, research is a multi-faceted and complex endeavor. Its outputs and outcomes vary, in particular by field, so measuring research impact can be challenging. In this paper, we reflect on the concept of field-weighting and discuss field-weighting methodologies. We study applications of field-weighting for Mendeley reads and present comparative analyses of field-weighted citation impact (FWCI) and field-weighted readership impact (FWRI). We see that there is a strong correlation between the number of papers cited and read per country. Overall, per subject area for the most prolific countries, FWCI and FWRI values tend to be close. Variations per country tend to hold true per field. FWRI appears to be a robust metric that can offer a useful complement to FWCI, in that it provides insights on a different part of the scholarly communications cycle.

Keywords: Research evaluation · Research performance
Research assessment · Metrics · Bibliometrics · Altmetrics · Indicators
Field-weighting

1 Introduction

The measurement of research impact is a growing and dynamic area, which is becoming more and more important. National assessment exercises such as the UK REF (Research Excellence Framework) and the Australian ERA (Excellence in Research for Australia) use research impact metrics, such as the h-index [1] and the journal Impact Factor [2] as part of their assessment criteria to distribute billions of pounds/dollars of research funding. Couple that with the growing number of global and national university rankings and the way that funding bodies assess funding applications, and the growing use of research impact metrics is clear.

© Springer Nature Singapore Pte Ltd. 2018
M. Erdt et al. (Eds.): AROSIM 2018, CCIS 856, pp. 96–104, 2018.
https://doi.org/10.1007/978-981-13-1053-9_8

One of the most widely used traditional metrics, is the number of citations an entity has received. Be it a paper, journal, researcher, group of researchers, or an institution, citation counts give an indication of how many times the research was referred to in other scientific publications. As with all metrics, the citation count has some weaknesses, one of them being that you cannot compare the citation counts of documents from different subject fields. Studies have shown that fields such as mathematics are likely to be cited far less than fields such as biochemistry & molecular biology [3]. Even when comparing papers within the same field, you must also take document age into consideration. For example, when published in the same field, does a paper published in 2005 with 40 citations have a bigger impact than one published in 2014 with 10 citations? The 2005 paper has had 9 additional years during which to build up citations, so it is not a fair comparison. Some researchers have employed PageRank algorithms in an attempt to overcome the weaknesses of citation counts and provide an alternative indicator to represent the academic influence of scientific papers [4–6].

Another way to compare outputs in fields of differing citation densities and ages is to use field normalization. The key role of such normalized indicators is to remove the effect that variables such as the age and subject area of a document have on a citation analysis, so that you can freely compare different documents against each other. At an article level, there are 2 commonly used normalization methods - normalization based on average citation counts and based on highly cited publications [7]. To calculate normalization metric for an article based on average citation counts, you need to know the expected number of citations for that publication. The expected number of citations for a paper is defined by calculating the average number of citations that similar document types in the same subject areas and year have received. For normalization calculations done using Scopus data, the subject areas are defined as the 334 ASJCs (All Science Journal Classifications). Examples of a normalization metric based on average citation counts include the field-weighted citation impact (FWCI) and the category normalized citation impact (CNCI) [8]. Both indicators are calculated at paper level and can be aggregated to provide normalized values for an author, group of authors, institution etc., by calculating the average for all normalized values for all documents in the set. For journals or serial titles, dedicated normalized indicators have been created to allow cross discipline comparisons. One example is SNIP (source normalized impact per paper), which measures a journal's contextual citation impact. Using Scopus data, it is calculated annually by the Centre for Science and Technology Studies (CWTS) at Leiden University [9].

Another form of normalization is based on highly cited publications. Here, field-dependent thresholds are used to determine if a publication is deemed to be highly cited [7]. The CiteScore Percentile metric is an example of a journal metric that uses such a methodology, which allows comparison between titles in different subject areas [10].

With the appearance of PlumX metrics and Altmetric.com, alternative metrics, or altmetrics, so named to differentiate themselves from the traditional metrics of citation and document indicators [11], are available to compliment traditional metrics. In PlumX for example, the metrics available include usage (clicks, views, downloads, library holdings, video plays), captures (bookmarks, favorites, reference manager saves), mentions (blog posts, news mentions, comments, reviews, Wikipedia mentions), social media (tweets, +1s, likes, shares) and citations (citation indexes, patent citations, clinical citations, policy citations) [12]. Many studies have been conducted on the advantages and disadvantages of altmetrics and potential correlations to citation counts [13, 14] and if researchers have an appetite and willingness to use more metrics such as usage data [15]. Whilst this paper does not go into the pros and cons of alternative metrics, we realize a growing willingness to use such indicators in the available "basket of metrics". In a recent randomized study of data collected from Impact Story, Mendeley readership counts provided the most metrics [16]. As such we wanted to investigate if it was possible to create a normalized version of the Mendeley readership.

2   Methodology

2.1   Data Sources

Scopus (https://www.scopus.com). Scopus is the world's largest abstract and citation database of peer-reviewed literature, delivering a comprehensive overview of global research output in the fields of science, technology, medicine, social science, and arts & humanities. Scopus includes abstracts and citation information from more than 70 million records including peer-reviewed journals, books and conference papers. Content coverage of peer-reviewed literature in Scopus includes research articles in 22,800 peer-reviewed journals published by over 5,000 publishers. Scopus covers approximately 6,400 titles from North-America, 11,800 from Europe, 2,500 from Asia-Pacific, and 1,500 from Latin-America and Africa.

Mendeley (https://www.mendeley.com). Mendeley is a free reference manager and academic social network that helps researchers organize their research, collaborate with others online, and discover the latest research. It has millions of users, including not only students, post-doctoral researchers, professors/lecturers, and other academic researchers but also commercial R&D professionals, government/NGO researchers, and other professionals. Mendeley can capture all types of reference information. However, in this paper we only analyze the publications that can be found in Scopus.

2.2   Indicators

Citations. Citations are formal references to earlier work made in an article, frequently to other journal articles. A citation is used to credit the originator of an idea or finding and is frequently used to indicate that the earlier work supports the claims of the work citing it. The number of citations received by an article from subsequently-published articles has been used as a proxy of the quality or importance of the reported research.

Readers. Readers refer to the number of Mendeley users who have added a particular article into their personal library. Mendeley reads are the counts of such events showing an early indicator of the impact a work has, both on receptivity of other authors within or beyond the same field as the work's author as well as non-authors such as clinicians, policymakers, funders, and students. Studies have also shown that readership numbers are a good early indication of future citation impact (Thelwall and Sud, 2015).

2.3 Field-Weighting

In order to study and compare the different indicators of impact previously defined, we computed field-weighted indicators. Field-weighted indicators allow the comparison of different publications because they take into account differences due to different document types, publication years, and subject areas. The same methodology has been applied to citations (FWCI, field-weighted citation impact) and readers (FWRI, field-weighted readership impact). They are indicators of mean citation/readership impact, and compare the actual number of citations/readers received by an article with the expected number of citations/readers for articles of the same document type, publication year, and field. If the article is classified in two or more subject areas, the harmonic mean of the actual and expected citation rates is used. The indicator is therefore always defined with reference to a global baseline of 1.0 and intrinsically accounts for differences in citation/reader accrual over time, differences in citation/reader rates for different document types (reviews typically attract more citations than research articles, for example), and differences in citation/reader rates across subject fields.

3 Results and Discussions

Figure 1 shows that at country level, there is a strong correlation between the number of papers cited and read per country. This resonates with the findings from other Mendeley based analyses [16–18]. Among prolific countries, established research nations such as the US and European countries tend to have relatively more publications read than cited, compared to the trend line. Conversely, several emerging research nations, such as BRIC and some Asian countries, tend to have relatively fewer publications read than cited, compared to the trend line. To some extent, this might be explained by the global distribution of Mendeley readers: as shown in Fig. 2, BRIC and some Asian countries tend to have relatively large number of publications relative to their number of Mendeley readers.

The correlation between field-weighted citation impact and field-weighted readership impact (thereby removing any size effects, as well as any specialization effects) is lower but still strong as demonstrated in Fig. 3, and holds regardless of publication output. Interestingly, several prolific countries (Brazil, India, Japan, Russia) that have relatively fewer publications read than cited compared to the trend line, have a rela-

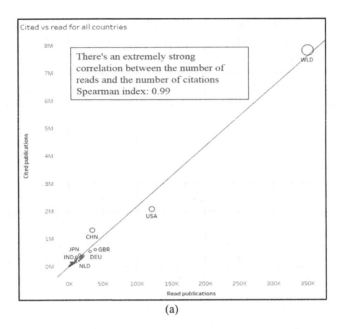

(a)

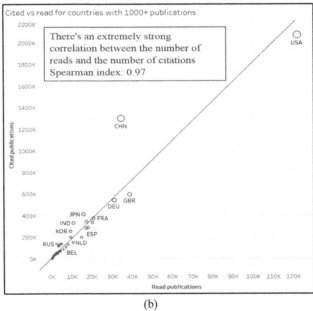

(b)

Fig. 1. Number of publications (size of circle), number of publications cited (y-axis), number of publications read (x-axis); for world and per country (a), for countries with 1,000+ publications (b); 2011–2015; sources: Scopus & Mendeley.

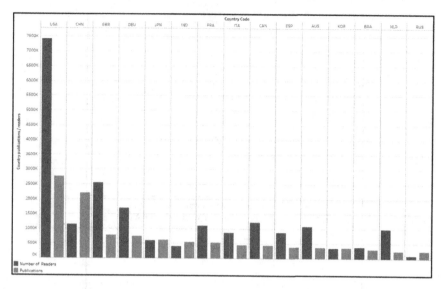

Fig. 2. Number of publications (orange), number of readers (blue); per country with 50,000+ publications; 2011–2015; (Color figure online) sources: Scopus & Mendeley.

tively higher than expected FWRI in relation to their FWCI. Conversely, China and several prolific established research nations have a relatively lower than expected FWRI in relation to their FWCI.

Figure 4 shows that for the world and for the most prolific countries, overall and per subject area, the FWCI and FWRI values tend to be close to the trend line, with a very slight FWRI advantage. Nevertheless, variations per country observed overall tend to hold true per OECD field. For instance, China's FWCI advantage holds true for each of the six OECD subject categories.

Looking at absolute comparisons of FWCI vs. FWRI as in Fig. 5 however reveals different patterns. While global values are close to 1 overall and for most subject areas, in most of fields except for the Humanities, there is a very slight FWRI advantage. The FWRI is however notably higher than the FWCI in the Agricultural Sciences and especially in the Social Sciences, in which the FWRI is 20% higher than the FWCI. There are also differences per country: in absolute terms, FWRI values tend to be higher than FWCI values across fields for prolific established western nations like the USA, the UK, Germany, and France, while the reverse is true for prolific Asian countries like China, Japan, or India.

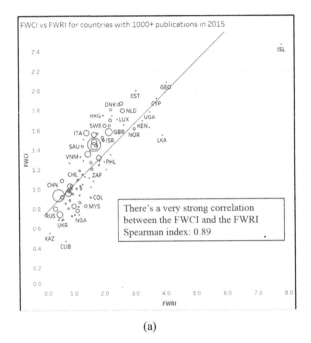

(a)

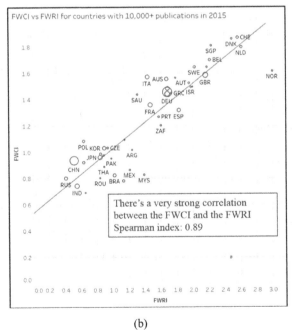

(b)

Fig. 3. Number of publications (size of circle), field-weighted citation impact (y-axis), field-weighted readership impact (x-axis); per country with 1,000+ publications (a), per country with 10,000+ publications (b); 2015; sources: Scopus & Mendeley.

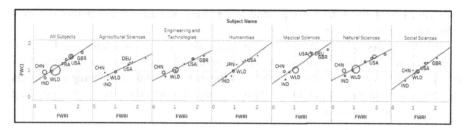

Fig. 4. Number of publications (size of circle), field-weighted citation impact (y-axis), field-weighted readership impact (x-axis); for world and per country with 100,000+ publications, overall and per OECD field; 2015; sources: Scopus & Mendeley.

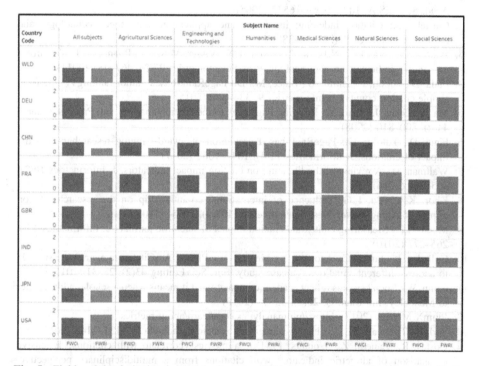

Fig. 5. Field-weighted citation impact (red) versus field-weighted readership impact (teal); for world and per country with 100,000+ publications; 2015; (Color figure online) sources: Scopus & Mendeley.

4 Conclusions

Our analyses show there is a strong correlation between number of papers cited and read per country, which doesn't appear to be a function of size. Specific regional patterns may be influenced by the geographic distribution of Mendeley readers. The correlation between field-weighted citation impact and field-weighted readership impact is lower but still strong. Overall and per subject area, and for most prolific countries, FWCI and FWRI values tend to be close. Variations per country observed overall tend to hold true per field.

FWRI appears to be a robust metric that can offer a useful complement to FWCI, in that it provides insights on a different part of the scholarly communications cycle. More detailed analyses are welcome to further test the metrics at different aggregation levels. It would also be interesting to see how it compares to other indicator types (e.g. downloads, views, altmetrics).

References

1. Hirsch, J.E.: An index to quantify an individual's s scientific research output. Proc. Natl. Acad. Sci. U.S.A. **102**, 16569–16572 (2005)
2. Garfield, E.: Citation indexing: its theory and application in science, technology, and humanities. Libr. Q. **50**(3), 384–385 (1979)
3. Waltman, L., van Eck, N.J., van Leeuwen, T.N., Visser, M.S., van Raan, A.F.J.: Towards a new crown indicator: an empirical analysis. Scientometrics **87**(3), 467–481 (2011)
4. Fiala, D., Šubelj, L., Žitnik, S., Bajec, M.: Do PageRank-based author rankings outperform simple citation counts? J. Informetrics **9**, 334–348 (2015)
5. Ma, N., Guan, J., Zhao, Y.: Bringing PageRank to the citation analysis. Inf. Process. Manag. **44**(2), 800–810 (2008)
6. Fiala, D., Tutoky, G.: PageRank-based prediction of award-winning researchers and the impact of citations. J. Informetrics **11**(4), 1044–1068 (2017)
7. Waltman, L.: A review of the literature on citation impact indicators. J. Informetrics **10**(2), 365–391 (2016)
8. Khor, K.A., Yu, L.G.: Influence of international co-authorship on the research citation impact of young universities. Scientometrics **107**(3), 1095–1110 (2016)
9. Moed, H.F.: Measuring contextual citation impact of scientific journals. J. Informetrics **4**(3), 265–277 (2010)
10. Colledge, L., James, C., Azoulay, N., Meester, W., Plume, A.: CiteScore metrics are suitable to address different situations – a case study. Eur. Sci. Editing **43**(2), 27–31 (2017)
11. Gunn, W.: Social signals reflect academic impact: what it means when a scholar adds a paper to Mendeley. Inf. Stan. Q. **25**(2), 33–39 (2013)
12. PlumX Metrics (2017). https://plumanalytics.com/learn/about-metrics/
13. Sud, P., Thelwall, M.: Evaluating altmetrics. Scientometrics **98**(2), 1131–1143 (2014)
14. Costas, R., Zahedi, Z., Wouters, P.: Do "altmetrics" correlate with citations? Extensive comparison of altmetric indicators with citations from a multidisciplinary perspective. J. Assoc. Inf. Sci. Technol. **66**(10), 2003–2019 (2015)
15. Colledge, L., James, C.: A "basket of metrics"—the best support for understanding journal merit. Eur. Sci. Editing **41**(3), 61–65 (2015)
16. Zahedi, Z., Costas, R., Wouters, P.: How well developed are altmetrics? A cross-disciplinary analysis of the presence of "alternative metrics" in scientific publications. Scientometrics **101** (2), 1491–1513 (2014)
17. Maflahi, N., Thelwall, M.: When are readership counts as useful as citation counts? Scopus versus Mendeley for LIS journals. J. Assoc. Inf. Sci. Technol. **67**(1), 191–199 (2016)
18. Mohammadi, E., Thelwall, M., Haustein, S., Larivière, V.: Who reads research articles? An altmetrics analysis of Mendeley user categories. J. Assoc. Inf. Sci. Technol. **66**(9), 1832–1846 (2015)

A Comparative Investigation on Citation Counts and Altmetrics Between Papers Authored by Top Universities and Companies in the Research Field of Artificial Intelligence

Feiheng Luo, Han Zheng⬤, Mojisola Erdt⬤,
Aravind Sesagiri Raamkumar$^{(\boxtimes)}$⬤, and Yin-Leng Theng

Wee Kim Wee School of Communication and Information,
Nanyang Technological University, Singapore 637718, Singapore
{fhluo,zhenghan,mojisola.erdt,aravind0002,
tyltheng}@ntu.edu.sg

Abstract. Artificial Intelligence is currently a popular research field. With the development of deep learning techniques, researchers in this area have achieved impressive results in a variety of tasks. In this initial study, we explored scientific papers in Artificial Intelligence, making comparisons between papers authored by the top universities and companies from the dual perspectives of bibliometrics and altmetrics. We selected publication venues according to the venue rankings provided by Google Scholar and Scopus, and retrieved related papers along with their citation counts from Scopus. Altmetrics such as Altmetric Attention Scores and Mendeley reader counts were collected from Altmetric.com and PlumX. Top universities and companies were identified, and the retrieved papers were classified into three groups accordingly: university-authored papers, company-authored papers, and co-authored papers. Comparative results showed that university-authored papers received slightly higher citation counts than company-authored papers, while company-authored papers gained considerably more attention online. In addition, when we focused on the most impactful papers, i.e., the papers with the highest numbers of citation counts, and the papers with the largest amount of online attention, companies seemed to make a larger contribution by publishing more impactful papers than universities.

Keywords: Citation analysis · Altmetrics · Industrial research
Academic research

1 Introduction

Artificial Intelligence is currently a widely discussed topic, not only in scientific research community but also in daily life, especially after a significant milestone was accomplished when a computer program called AlphaGo defeated a human professional player in a five-game Go match in 2016 [1]. Both academia and industry have made a lot of efforts in this research area, and achieved impressive results in a number of tasks including computer Go program, machine translation, speech recognition,

© Springer Nature Singapore Pte Ltd. 2018
M. Erdt et al. (Eds.): AROSIM 2018, CCIS 856, pp. 105–114, 2018.
https://doi.org/10.1007/978-981-13-1053-9_9

object detection, to name a few [1–7]. Typically, deep learning is one of the most popular and the most impactful techniques employed in these studies. However, deep learning requires a huge amount of training data and powerful computing resources [1, 8, 9]. Compared to universities, influential companies such as Google could have more advantages on these two requirements. Hence, in the current study, we are interested in evaluating the research outputs produced by academia and industry in the research field of Artificial Intelligence.

In this paper, we conducted a comparative investigation on citation counts and altmetrics between scientific papers published by the top academic universities and industrial companies in the research fields of Artificial Intelligence. From Scopus, we extracted papers published in the relevant top publication venues that were determined through Google Scholar and Scopus, for the years 2015 and 2016. Amongst these papers, we identified the top universities and companies. Comparisons of bibliometrics and altmetrics were made among three groups of papers: university-authored papers, company-authored papers, and co-authored papers. Comparative results showed that only a small difference in citation counts among the three groups of papers was observed, while company-authored papers received much more attention online than university-authored papers.

In the following sections, we will describe data collection and data processing procedures, followed by results of a variety of comparisons. Finally, we will conclude this paper with a discussion.

2 Data

We considered the data sources Scopus, Google Scholar, Altmetric.com and PlumX. Details of data collection and data processing are presented in this section. Data collection was completed in August 2017.

2.1 Publication Venue Selection

Artificial Intelligence research field is the main focus of this study. First we identified related publication venues. Different from other research fields, Artificial Intelligence has given conference papers an equal or even higher value than journal papers, and thus both conference papers and journal papers were considered in current study. As one of the most prevailing academic search engines, Google Scholar provides top publication venue lists for different research fields. For Artificial Intelligence, Google Scholar lists the top 20 publication venues consisting of journals and conference proceedings based on the h5-index, the h-index for articles published in the last five complete years [10]. In the top 20 publication venues, we excluded arXiv venues since papers uploaded on arXiv are mostly preprints. We also removed venues in which the paper DOI was not available since paper DOIs are required for fetching altmetric data. Finally, we obtained a list of top 10 venues from Google Scholar.

We also selected publication venues from Scopus. However, the research fields in Scopus have been classified into broad categories, and Artificial Intelligence is classified under Computer Science. Thus, we selected the top 10 journals and the top 10 conference proceedings based on the Scopus CiteScore which measures average citations received per document published in the venue [11]. We also removed venues that did not contain the paper DOI.

Since the h5-index is used in ranking the venues on Google Scholar, the publication venues which publish more papers are benefitted by it. But on the other hand, the Scopus CiteScore played a similar role as journal impact factor. As a result, by combining both Google Scholar and Scopus, a total of 30 publication venues without overlaps were identified as the venues related to Artificial Intelligence in the current study.

Subsequently, we retrieved all research papers published in these 30 venues in 2015 and 2016 from Scopus. There were 14,647 retrieved papers in total, and the exported information of each paper included title, DOI and affiliations. Table 1 gives a summary of the selected venues.

Table 1. Summary of selected venues.

S/N	Venue	Source	# of papers
1	Expert Systems with Applications	GS	1,393
2	IEEE Transactions on Neural Networks and Learning Systems	GS	542
3	IEEE Transactions on Fuzzy Systems	GS	330
4	Knowledge-Based Systems	GS	68
5	Applied Soft Computing	GS	1,198
6	Neurocomputing	GS	2,901
7	Neural Networks	GS	93
8	Engineering Applications of Artificial Intelligence	GS	385
9	Neural Computing and Applications	GS	321
10	Robotics and Autonomous Systems	GS	249
11	IEEE Transactions on Pattern Analysis and Machine Intelligence	Scopus	386
12	IEEE Transactions on Evolutionary Computation	Scopus	124
13	Wiley Interdisciplinary Reviews: Computational Molecular Science	Scopus	32
14	ACM Computing Surveys	Scopus	111
15	International Journal of Computer Vision	Scopus	150
16	IEEE Communications Magazine	Scopus	526
17	IEEE Internet of Things Journal	Scopus	162
18	IEEE Journal on Selected Areas in Communications	Scopus	477
19	IEEE Wireless Communications	Scopus	181
20	IEEE Signal Processing Magazine	Scopus	125
21	Proceedings of the IEEE Computer Society Conference on Computer Vision and Pattern Recognition	Scopus	1,245

Table 1. (*continued*)

S/N	Venue	Source	# of papers
22	Proceedings of the IEEE International Conference on Computer Vision	Scopus	600
23	International Conference on Architectural Support for Programming Languages and Operating Systems	Scopus	102
24	Proceedings of the ACM SIGKDD International Conference on Knowledge Discovery and Data Mining	Scopus	455
25	Proceedings of the Annual ACM Symposium on Theory of Computing	Scopus	185
26	Proceedings of the ACM SIGMOD International Conference on Management of Data	Scopus	390
27	Proceedings - International Conference on Software Engineering	Scopus	582
28	Proceedings of the Annual International Conference on Mobile Computing and Networking, MOBICOM	Scopus	223
29	Proceedings - IEEE INFOCOM	Scopus	1,004
30	Proceedings of the ACM SIGPLAN Conference on Programming Language Design and Implementation	Scopus	107

2.2   Prestigious Affiliation Identification

This study intends to explore the comparisons between papers authored by the academic institutes and papers authored by industry in terms of bibliometrics and altmetrics. Since the top publication venues were selected in the last subsection, we considered the prestigious affiliations, due to our assumption that less prestigious affiliations might publish fewer papers in those top venues.

In the 14,647 retrieved papers, an affiliation could be recorded in different names. For example, "Google" also had another recorded name as "Google Inc". We used string match to consolidate the affiliation names, so that every affiliation entry has a unique name. We sorted the affiliations by the number of affiliated papers, and identified the top 10 universities and the top 10 companies. These identified affiliations published the most papers in the selected venues. In general, academic universities are more engaged in the research community, and thus we additionally included the top 10 universities on the QS World University Rankings 2015 for Computer Science [12]. Since there was an overlap between the two lists of top 10 universities, we finally obtained 19 unique universities and 10 companies, which are summarized in Table 2.

Table 2. Summary of identified universities and companies.

Affiliation	Type	# of papers
MIT	University	175
Stanford University	University	181
University of Oxford	University	47
Carnegie Mellon University	University	210
Harvard University	University	58
University of California, Berkeley	University	211
University of Cambridge	University	39
The Hong Kong University of Science and Technology	University	108
ETH Zurich - Swiss Federal Institute of Technology	University	119
Princeton University	University	75
Tsinghua University	University	450
Harbin Institute of Technology	University	229
Southeast University	University	198
Nanyang Technological University	University	203
Huazhong University of Science and Technology	University	205
Xidian University	University	194
Shanghai Jiao Tong University	University	210
Zhejiang University	University	180
National University of Singapore	University	173
Microsoft Research	Company	318
IBM Research	Company	136
Google	Company	121
Huawei Technologies	Company	88
NEC Laboratories America	Company	61
Ericsson Research	Company	59
Nokia	Company	59
Samsung Electronics	Company	57
Yahoo Labs	Company	51
Adobe Research	Company	49

2.3   Retrieval of Bibliometric and Altmetric Data

We collected citation counts of the retrieved papers using the Scopus API. Out of the total 14,647 papers, citation counts could be extracted for 14,474 papers (98.8%), while access errors were encountered when fetching citation counts of the remaining 173 papers. Please note that an available citation count simply meant that the citation count of the particular paper could be retrieved from Scopus via the API, but it did not mean the number was necessarily greater than zero.

We harvested altmetric data from two predominant sources: PlumX and Altmetric. com. Most of the papers had been indexed in PlumX, since PlumX and Scopus both belong to Elsevier. When we used Altmetric.com API, the number of papers with retrieved data was much lower. However, we found that if the preprint version of a paper was uploaded to arXiv, then the preprint version could be also indexed in Altmetric.com, which meant that a particular paper could have two different records on Altmetric.com: one for the preprint version and the other for the print version. It was also found that for some of the papers, only the preprint versions were indexed in Altmetric.com. Thus, we used full title match to obtain arXiv ids for all the papers if they were uploaded to arXiv, and fed their arXiv ids to the Altmetric.com API to retrieve altmetric data.

From PlumX, we collected four types of metrics: usage data (e.g., views count and downloads count), captures data (e.g., Mendeley readers count and bookmarks count), mentions data (e.g., blog mentions count), and social media metrics data (e.g., Facebook posts count and tweets count), all of which were the main indicators in PlumX. From Altmetric.com, we collected three metrics: Altmetric Attention Scores, Mendeley readers, and tweets. The Altmetric Attention Score is an indicator of the amount of attention the paper has received online [13]. It is a weighted count of the online attention a paper has received, such as references in Wikipedia and mentions in social networks. In addition, we also collected Mendeley readers count and tweets count, since both metrics were available for relatively more papers. For a certain paper, if it had two records in Altmetric.com, we selected the higher value for each metric.

3   Results

Comparisons of collected citation counts and altmetrics were made between papers published by the 19 identified universities and papers published by the 10 companies. In the comparisons, if there was an author affiliated with the identified university or the identified company, then the corresponding paper was defined as a university-authored paper or a company-authored paper. Additionally, a paper was defined as a co-authored paper, if its associated affiliations involved both the identified university and company. Comparison results are presented in this section.

3.1   Comparison of Citation Counts

Figure 1 compares citation counts among the three paper groups: university-authored papers, company-authored papers, and co-authored papers. The left subfigure of Fig. 1 shows citation counts in a boxplot in which the line in the box indicates the median of citation counts. According to the left subfigure, papers published by universities had slightly higher citation counts, but overall there was not much difference among the three groups. The right subfigure of Fig. 1 compares means of citation counts among the three paper groups, where the company-authored paper group obtains the highest mean of citation counts. From Fig. 1, the results revealed that in general, whether the paper was published by the university or by the company, it had a limited impact on the citation count. However, when considering highly cited papers, companies may have a larger contribution. By counting the number of highly cited papers which were defined as papers with at least 100 citations, it was validated that there were in total 20 highly cited papers, consisting of 10 company-authored papers, 9 university-authored papers, and 1 co-authored paper.

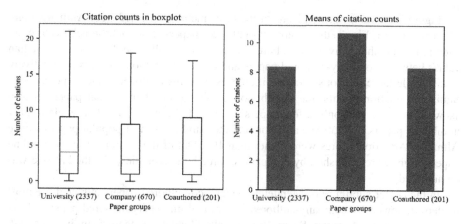

Fig. 1. Comparison of citation counts among three paper groups.

3.2 Comparison of Altmetrics

Comparisons of altmetric data collected from PlumX are first presented. The four types of metrics collected from PlumX were sparse. While capture data was available for most of the papers, usage data was available for around half of the papers, and for mentions data and social media metrics data, the percentages of papers with non-null records were lower than 10%. According to Fig. 2, it is observed that co-authored papers achieved the best performance for both usage and captures data. In addition, there is no major difference between the university-authored paper group and the company-authored paper group in the left subfigure of usage data, but company-authored papers indeed outperform university-authored papers in the right subfigure of captures data. Regarding mentions and social media metrics data, the comparison results were insufficiently representative, since only a very small number of papers had non-null records, but we found that the papers with the highest values of mentions and social media metrics data were published by companies.

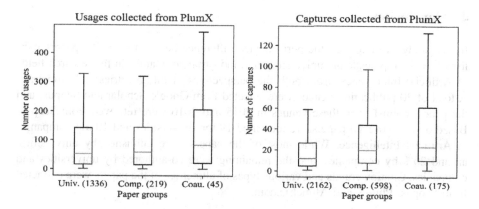

Fig. 2. Comparisons of usage and captures collected from PlumX.

Figure 3 illustrates comparisons of the three metrics collected from Altmetric.com. We can see in Fig. 3 that company-authored papers consistently achieved more attention online than university-authored papers. Specifically, the differences of median values between company-authored papers and university-authored papers are relatively small, while there are considerable differences in terms of upper range of percentiles. Similar to citation counts, the distribution of altmetrics of individual papers is also skewed, meaning that only a few papers received high attention online while a large number of papers did not. We also counted the number of most popular papers whose Altmetric Attention Scores were higher than 100. Out of the 12 most discussed-online papers, eight were published by companies, three by universities, and the last one was co-authored.

In the center subfigure of Mendeley readers, co-authored papers attracted the most readers, followed by company-authored papers and university-authored papers. However, the Altmetric Attention Score does not take Mendeley readers count into account, which could be a reason why the comparative pattern in the center subfigure is different from the patterns in the other two subfigures.

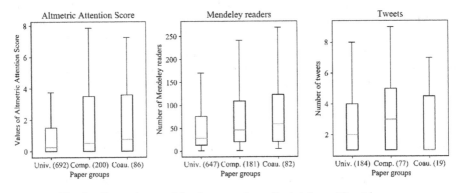

Fig. 3. Comparisons of the three metrics collected from Altmetric.com.

4 Conclusion

In this study, we explored the performance difference between scientific papers published by the top academic universities and industrial companies in the research fields of Artificial Intelligence, from both the perspectives of bibliometrics and altmetrics. A total of 30 publication venues were selected from Google Scholar and Scopus, and the papers published in these venues in 2015 and 2016 were retrieved from Scopus. Based on the retrieved papers, we identified 19 top universities and 10 top companies in Artificial Intelligence. While most of the papers were published by universities, around 20% by companies, and the remaining were co-authored by universities and companies. Citation counts and various types of altmetrics of the papers were extracted from Scopus, PlumX and Altmetric.com.

Comparisons of citation counts among the three groups of papers showed that university-authored papers received slightly higher citation counts but the difference was relatively small. On the contrary, comparisons of altmetrics revealed that company-authored papers and co-authored papers obtained considerably more attention online than university-authored papers. However, the comparison results could be affected by the unbalanced numbers of the top universities and companies. In addition, some important venues in Artificial Intelligence such as the International Conference on Machine Learning (ICML) were not included in the current study due to the lack of DOIs of the papers published in these venues.

Most Internet users might not be scientific researchers, and thanks to the diverse products and services provided by companies, people would be more familiar with the large, well-known companies and would pay more attention to them online. In addition, research conducted by companies could be more product-oriented and closer to daily life. These factors may motivate people to discuss more about papers authored by companies online. However, regarding paper citation count, it depends essentially on the academic quality, rigor and novelty of the paper. Hence, there was not much difference in citation counts between university-authored papers and company-authored papers. This is simply our intuitive explanation, while a deep understanding should be acquired through a series of well-designed surveys and interviews.

Although there is a consistent growth of studies on altmetrics, the relationship between bibliometrics and altmetrics has not been fully revealed [14]. In our study, we focus on the papers published by industries, since this typical category of papers would probably have advantages in receiving more attention online. In future work, we plan a thorough investigation of the relationship between citation counts and altmetrics for company-authored papers. A larger set of company-authored papers would be collected and investigated on the content of their altmetrics. We would also look into the qualitative differences between altmetrics for company-authored papers and university-authored papers.

Acknowledgement. This research is supported by the National Research Foundation, Prime Minister's Office, Singapore under its Science of Research, Innovation and Enterprise programme (SRIE Award No. NRF2014-NRF-SRIE001-019). We also thank Altmetric.com for providing access to Fetch API.

References

1. Silver, D., Huang, A., Maddison, C.J., Guez, A., Sifre, L., Van Den Driessche, G., Dieleman, S., et al.: Mastering the game of Go with deep neural networks and tree search. Nature **529**(7587), 484–489 (2016)
2. Cho, K., Van Merriënboer, B., Gulcehre, C., Bahdanau, D., Bougares, F., Schwenk, H., Bengio, Y.: Learning phrase representations using RNN encoder-decoder for statistical machine translation. arXiv preprint arXiv:1406.1078 (2014)
3. Vaswani, A., Shazeer, N., Parmar, N., Uszkoreit, J., Jones, L., Gomez, A.N., Kaiser, L., Polosukhin, I.: Attention is all you need. arXiv preprint arXiv:1706.03762 (2017)

4. He, K., Zhang, X., Ren, S., Sun, J.: Deep residual learning for image recognition. In: Proceedings of the IEEE Conference on Computer Vision and Pattern Recognition, pp. 770–778 (2016)
5. Ren, S., He, K., Girshick, R., Sun, J.: Faster R-CNN: towards real-time object detection with region proposal networks. In: Advances in Neural Information Processing Systems, pp. 91–99 (2015)
6. Graves, A., Mohamed, A.R., Hinton, G.: Speech recognition with deep recurrent neural networks. In: Proceedings of the IEEE International Conference on Acoustics, Speech and Signal Processing, pp. 6645–6649 (2013)
7. Chorowski, J.K., Bahdanau, D., Serdyuk, D., Cho, K., Bengio, Y.: Attention-based models for speech recognition. In: Advances in Neural Information Processing Systems, pp. 577–585 (2015)
8. LeCun, Y., Bengio, Y., Hinton, G.: Deep learning. Nature **521**(7553), 436–444 (2015)
9. Schmidhuber, J.: Deep learning in neural networks: an overview. Neural Netw. **61**, 85–117 (2015)
10. Google Scholar top publications in Artificial Intelligence. https://scholar.google.com.sg/citations?view_op=top_venues&hl=en&vq=eng_artificialintelligence. Accessed 6 Oct 2017
11. Scopus list of sources. https://www.scopus.com/source/browse.url. Accessed 6 Oct 2017
12. QS World University Rankings by Subject 2015 - Computer Science & Information Systems. https://www.topuniversities.com/university-rankings/university-subject-rankings/2015/computer-science-information-systems. Accessed 6 Oct 2017
13. Altmetric Attention Score. https://www.altmetric.com/blog/the-altmetric-score-is-now-the-altmetric-attention-score/. Accessed 6 Oct 2017
14. Erdt, M., Nagarajan, A., Sin, S.C.J., Theng, Y.L.: Altmetrics: an analysis of the state-of-the-art in measuring research impact on social media. Scientometrics **109**(2), 1117–1166 (2016)

Scientometric Analysis of Research Performance of African Countries in Selected Subjects Within the Field of Science and Technology

Yusuff Utieyineshola$^{(\boxtimes)}$ (iD)

National Centre for Technology Management (NACETEM), Abuja, Nigeria
yusuffshola@yahoo.com

Abstract. This paper assesses the performance of African countries in selected field of Science and Technology (S&T) over the last twenty years. The purpose is to determine the readiness of these countries in aligning to the strategic direction set by African Union (AU 2063) agenda. The AU 2063 aims to emplace a paradigm shift from the current structure where its members' dependence on natural resources to drive their economies to one that is knowledge-based. It thus set pillars for archiving this feat and they include; building and/or upgrading research infrastructures; enhancing professional and technical competencies; promoting entrepreneurship and innovation; and providing an enabling environment for STI development in the African continent. Data used for the study were retrieved from the SCImago database which comprises a total of seven subject areas cutting across 126 subject categories. In SCImago database, information was also searched on S&T performances with respect to publications in the World and Africa, over the last 20 years period starting from 1996–2015. Microsoft Excel was used to analyse the data collected. Results were presented in tables and figures on the top 10 most productive African countries in the field of S&T in all the seven selected subject areas. The paper suggested an intra-African collaborative effort between low and high performing countries in Africa as an option for developing the needed knowledge capacities for realising its regional developmental agenda (AU 2063).

Keywords: Scientometrics · Performance · Science and technology

1 Introduction

African leaders have seen the need to emplace the continent on a pedestal aimed towards self-reliance capable of promoting economies of its member states that is more sustainable and in tune with what is obtainable in the developing world. In 2014, to reaffirm its vision of "an integrated, prosperous and peaceful Africa, an Africa driven and managed by its own citizens and representing a dynamic force in the international arena", the African Union under its AU agenda 2063 recognized Science, Technology and Innovation (STI) as multi-functional tools and enablers for achieving continental development goals. Hence, the Science, Technology and Innovation Strategy for Africa

© Springer Nature Singapore Pte Ltd. 2018
M. Erdt et al. (Eds.): AROSIM 2018, CCIS 856, pp. 115–124, 2018.
https://doi.org/10.1007/978-981-13-1053-9_10

2024 (STISA-2024), was initiated. The STISA-2024 is the first of the ten-year incre-mental phasing strategies to respond to the demand for science, technology and innovation to impact across critical sectors such as agriculture, energy, environment, health, infrastructure development, mining, security and water among others. The strategy is firmly anchored on six distinct priority areas that contribute to the achievement of the AU vision. These priority areas are - *Eradication of Hunger and Achieving Food Security, Prevention and Control of Diseases, Communication (Physical and Intellectual Mobility), Protection of our Space, Live Together - Build the Society; Wealth Creation.* The strategy further defines four mutually reinforcing pillars which are considered as prerequisite conditions for its success. These pillars include - building and/or upgrading research infrastructures, enhancing professional and technical competencies, promoting entrepreneurship and innovation, and providing an enabling environment for STI development in the African continent. It anticipates that continental, regional and national programmes will be designed, implemented and synchronized to ensure that their strategic orientations and pillars are mutually rein-forcing, and achieve the envisaged developmental impact as effectively as possible.

Every positive-oriented society today needs skilled and talented individuals to generate new ideas, products, processes and commercial enterprises. Therefore, existing studies have shown that accessing performance on the basis of STI, African countries performance is rated poorly if measured on indicators as tertiary education institutions, intellectual property and innovativeness and productivity and competi-tiveness [1]. This position was also supported and explained by the United Nations Economic Commission for Africa (UNECA) in its African Science, Technology and Innovation Review 2013 report document. The review was done to assess STI status and performance in the African context with a view for describing the innovation ecosystem in Africa. It looks at the innovation value chain from the perspective of training and research and development; technology development, acquisition, use and application.

In the last decade, Africa has recorded an annual growth rate of about 15% in terms of enrolment rate in tertiary institutions while in 2008, the figure for Sub-Saharan African countries on this same indicator was only 6% which is lower when compared with statistics on other continents - Asia (26%), Latin America and the Caribbean (38%). Furthermore, in terms of researchers involved in R&D, Africa's performance is still relatively poor. For instance, in a survey conducted in 13 countries by African Science, Technology and Innovation Indicator Initiative (ASTII), the result shows that more than half of these countries have fewer than 1000 R&D researchers in total. However, only Gabon, Senegal and South Africa have more than 20% of their total R&D personnel with PhD qualifications while Mozambique and Kenya reported less than 2% for this indicator [2].

To ascertain these claims, different tool for assessing performance and productivity of a system such as scientometrics can be employed. Though there are other tools for assessing scientific production, however, scientometrics is very useful for this purpose. In the field of Science and Technology Studies (STS), scientometrics is a useful tool for measuring the scientific and technological performance of a knowledge system.

Scientometrics involves measurement of scientific publications using a method referred to as bibliometrics [3]. Scientometrics is restricted to the measurement of science communications, whereas bibliometrics is designed to deal with more general information processes [4]. Scientometrics is for science what econometrics is for economics [5]. The advent of Scientometrics journal in 1978 from a research unit in the Hungarian Academy of Science and Scientific conferences, led to the development of Scientometrics as a discipline [6]. They stated that it was developed around one core notion (citations) though the discipline can study (to some extent) many aspects of the dynamics of science and technology. The citation is not only important in scientometrics but provide a quantitative metrics for measuring research impact. Mingers and Leydesdorff [6] further buttress this position and state that "The act of citing another person's research provides the necessary linkages between people, ideas, journals and institutions to constitute an empirical field or network that can be analyzed quantitatively".

This paper seeks to use scientometrics to analyze research performance of African countries in selected subjects within the field of S&T.

2 Methodology

The research was designed based on the need to find the best approach that could lead to a logical route to addressing the objectives of the research. The focal objective of this study was specifically to study how African countries perform in S&T over the last twenty years (1996–2015). To this end, the research design approach upon which this study was built, rests on the previous research works of [7, 8] where in both cases, scientometrics analysis of publication output on S&T in India between 1989–2014 and 1996–2011 respectively were carried out by these scholars. Therefore, in this study, the sample population from the SCImago database used in this study comprises a total of seven subject areas having a total of 126 subject categories. They include; Agricultural and Biological Sciences (14), Biochemistry, Genetics and Molecular Biology (17), Chemical Engineering (9), Computer Science (13), Engineering (17), Material Science (9), Medicine (47). On SCImago database, information was also searched on S&T performances with respect to publications in the World, Africa, Asia and Nigeria over the last 20 years period starting from 1996–2015. But for the purpose of this paper, only data on Africa was used for analysis. On the SCImago database, the search query used was (Search = "World") AND (Year = 1996–2015); (Search = "Africa") AND (Year = 1996–2015); (Search = "Asia") AND (Year = 1996–2015); (Search = "Nigeria") AND (Year = 1996–2015), each done separately. The fata was retrieved between the fourth week of May 2017 and second week of June 2017. To analyse the data retrieved, Microsoft Excel was used. Results obtained were presented in tables on the top 10 most productive African countries in the field of S&T in all the seven selected subject areas.

3 Data Analysis and Discussion

The ten most productive countries in Africa in the field of Science are shown in Table 1. Their corresponding ranking in the world is also shown to reflect their position beyond the continent. South Africa is ranked first in Africa and 34th position in the world having produced 188,104 documents out of which 91.66% of it is citable. Ranked second in Africa is Nigeria with a world ranking of 52nd position, a wide margin from that of South Africa. Nigeria produced 59,372 documents between the years under review out of which 95.38% of them are citable. Nigeria, including South Africa and Tunisia, records a high percentage of self-citation in the region having more than 21%. Egypt supposed to stand at the second position in Africa considering its position of 42nd in the world; however, following the ranking list as obtained from the SCImago, the country was not included on the list. In terms of h-index, South Africa has the highest ranking followed by Kenya and Nigeria. Interestingly, in terms of citation per paper, Kenya recorded the highest score in this category despite its 6th position in Africa. This shows that despite the low volume of documents produced during the period under review, it was able to attract attention within the academic community. Overall, the Northern African countries prove to be very strong in the production of scientific knowledge in Africa having displayed more countries from the region according to the ranking. The performance of these North African countries may be as a result of their collaboration with their fellow countries in the Arab region such as Saudi Arabia and the Emirates where they also receive grants to promote their research activities. The overall performance of Africa as ranked in the world calls for improvement and the need to address those challenges that researchers in this part of the world face which directly impact on the number and quality of publications from the region.

Table 1. Top 10 most productive countries in Africa in the field of Science (all subject areas)

Rank (Africa)	Rank (World)	Country	Docs	Citable docs	% citable docs	Citations	Self-citations	% self-citations	Citations per document	H index
1	34	South Africa	188104	172424	91.66	2125927	454537	21.38	11.3	320
2	52	Nigeria	59372	56630	95.38	334059	72718	21.77	5.63	131
3	53	Tunisia	58769	55904	95.12	342429	73636	21.50	5.83	123
4	55	Algeria	42456	41544	97.85	215922	43297	20.05	5.09	106
5	56	Morocco	40737	38371	94.19	279731	51031	18.24	6.87	129
6	67	Kenya	24458	22347	91.37	379560	57594	15.17	15.52	179
7	78	Ethiopia	13363	12625	94.48	118656	24840	20.93	8.88	101
8	84	Tanzania	11964	11140	93.11	170144	25866	15.20	14.22	122
9	86	Ghana	11543	10578	91.64	111205	13874	12.48	9.63	105
10	87	Uganda	11528	10599	91.94	171367	26995	15.75	14.87	128

*Egypt ranked 42 in the World but it was not included the ranking list in Africa as obtained from source

Source: SCImago, Author analysis, 2017

In the field of Agricultural & Biological Science, South Africa and Nigeria still maintained the top two positions in Africa. Nigeria has the highest percentage of citable document (99.51%) as shown in Table 2 and is closely followed by Ethiopia which records 99.10%. South Africa has the highest case of self-citation (28.75%) followed by Ethiopia (25.37%) and Nigeria (24.34%). In terms of citation per document, Kenya tops this section having recorded 13.9% citations per document produced. Kenya has also performed well as indicated by the h-index having 103 behind South Africa which is ranked the first position in the field of Agricultural & Biological Science.

Table 2. Top 10 most productive countries in Africa in Agricultural & Biological Science

Rank	Country	Documents	Citable documents	% citable documents	Citations	Self-citations	% self citations	Citations per document	H index
1	South Africa	34375	33575	97.67	444511	127778	28.75	12.93	165
2	Nigeria	14339	14269	99.51	82412	20055	24.34	5.75	69
3	Kenya	8053	7894	98.03	111942	20305	18.14	13.9	103
4	Tunisia	6427	6341	98.66	58160	13936	23.96	9.05	76
5	Ethiopia	4223	4185	99.10	34820	8834	25.37	8.25	58
6	Morocco	3392	3347	98.67	40130	5532	13.79	11.83	67
7	Tanzania	3029	2965	97.89	34235	5377	15.71	11.3	65
8	Algeria	2955	2914	98.61	17994	3433	19.08	6.09	46
9	Cameroon	2845	2818	99.05	27540	5451	19.79	9.68	55
10	Uganda	2646	2598	98.19	28944	5335	18.43	10.94	59

Source: SCImago, Author analysis, 2017

According to the data presented in Table 3, recording a total of 18,946 in terms of document produced, South Africa ranked the highest position within the field of Biochemistry, Genetics and Molecular Biology between the periods under review. In terms of quality of documents produced, there is a fair distribution in the percentage of citable documents across all the 10 countries between the ranges of 98.74% being the highest from Nigeria and to the lowest coming from South Africa (96.57%). With 20.06%, Nigeria tops other countries in the case of self-citations. This is not to its credit as citations should be assessed from other researcher and not a function of authors citing their own work so as to boost its citation counts. The case of self-citation in Nigeria calls for concern if considering that despite being ranked 2nd in the continent in terms of number of documents produced, it has the lowest (7.64) citation per document and highest percentage of self-citation so far. Overall, South Africa's productivity in this subject field can be said to be balanced, having received the highest h-index in this category, over a double figure for the rest of the countries in the ranking.

Table 3. Top 10 most productive countries in Africa in Biochemistry, Genetics and Molecular Biology

Rank	Country	Docs	Citable documents	% citable documents	Citations	Self-citations	% self citations	Citations per document	H index
1	South Africa	18946	18297	96.57	327073	59190	18.10	17.26	162
2	Nigeria	6344	6264	98.74	48489	9728	20.06	7.64	68
3	Tunisia	6243	6113	97.92	70516	12936	18.34	11.3	84

(*continued*)

Table 3. (*continued*)

Rank	Country	Docs	Citable documents	% citable documents	Citations	Self-citations	% self citations	Citations per document	H index
4	Morocco	3104	3047	98.16	42632	4955	11.62	13.73	78
5	Kenya	2823	2764	97.91	47119	5915	12.55	16.69	81
6	Algeria	2168	2133	98.39	22552	2805	12.44	10.4	57
7	Cameroon	1477	1451	98.24	17555	3257	18.55	11.89	49
8	Ethiopia	1386	1371	98.92	14208	2131	15.00	10.25	47
9	Uganda	1190	1171	98.40	15938	2537	15.92	13.39	53
10	Tanzania	1105	1091	98.73	15126	1723	11.39	13.69	54

Source: SCImago, Author analysis, 2017

Research in the field of Chemical Engineering shows that South Africa tops the ranking list of publications in Africa (Table 4). Unlike in the other subject categories considered earlier, there is a departure from the usual in the percentage of citable documents produced where Sudan has 100% of its documents citable. Cameroon recorded the highest percentage of self-citation of 19.65% followed by South Africa. In terms of citations per documents, Morocco recorded the highest figure in the region. Chemical Engineering is an important field that plays a significant role in the production of chemicals for industries alike, Africa's research in this direction is commendable.

Table 4. Top 10 most productive countries in Africa in Chemical Engineering

Rank	Country	Documents	Citable documents	% citable documents	Citations	Self-citations	% self-citations	Citations per document	H-index
1	South Africa	4993	4928	98.70	69063	12622	18.28	13.83	90
2	Algeria	2649	2624	99.06	24591	3777	15.36	9.28	58
3	Tunisia	2576	2542	98.68	28698	4956	17.27	11.14	60
4	Nigeria	1635	1624	99.33	14886	2706	18.18	9.1	57
5	Morocco	1463	1453	99.32	25609	4210	16.44	17.5	68
6	Libya	288	286	99.31	1338	85	6.35	4.65	17
7	Cameroon	223	222	99.55	1588	312	19.65	7.12	20
8	Kenya	146	142	97.26	2789	95	3.41	19.1	22
9	Sudan	142	142	100.00	882	72	8.16	6.21	17
10	Ghana	131	125	95.42	951	66	6.94	7.26	14

Source: SCImago, Author analysis, 2017

Computer Science as a field of Science is very important in the world today. Virtually all human activities are dependent on one form of technology or the other. Over the years, Asian countries have built capacities and enforce their superiority in the field of Information and Communication Technology (ICT) over other developing countries. A look at figures in Table 5 shows that Africa's productivity in this subject area is still dominated by South Africa with a total of 10,644 documents produced. The North African countries appear to be more formidable in this subject field having displaced Nigeria to the 5th position in the ranking. In terms of citations per paper,

Table 5. Top 10 most productive countries in Africa in Computer Science

Rank	Country	Documents	Citable documents	% citable documents	Citations	Self-citations	% self-citations	Citations per document	H index
1	South Africa	10644	10456	98.23	50791	7854	15.46	4.77	77
2	Tunisia	9787	9666	98.76	23571	6161	26.14	2.41	49
3	Algeria	8168	8091	99.06	21501	4635	21.56	2.63	50
4	Morocco	4664	4618	99.01	10690	2451	22.93	2.29	36
5	Nigeria	2597	2574	99.11	4528	1160	25.62	1.74	24
6	Libya	492	487	98.98	1283	57	4.44	2.61	20
7	Ghana	375	370	98.67	747	94	12.58	1.99	15
8	Kenya	357	354	99.16	1280	166	12.97	3.59	19
9	Mauritius	335	330	98.51	635	76	11.97	1.9	12
10	Sudan	321	321	100.00	870	48	5.52	2.71	13

Source: SCImago, Author analysis, 2017

Kenya standing at the 8th position is closely ranked with South Africa having received 3.59 citations per document. Considering the total figure of documents produced, African researchers need to improve on their publication activity within this subject field since the relevance of Technology cross-cut all sectors of human endeavour today.

In today's world, Engineering concepts and applications have continued to react to the dynamics of the society. Either in Construction, Design, Machine fabrication or Industrial input, Engineering is an important field that is as old as humanity itself. South Africa is still the dominant country in this field has produced a total of 19,163 documents so far (Table 6). The North African countries (Tunisia and Algeria) are closely ranked after South Africa in terms of documents produced, citations per document and even in h-index received. Worthy of note here is that Tunisia and Algeria have more cases of self-citation in the region, an indication that is not favourable to the quality of publications.

Table 6. Top 10 most productive countries in Africa in Engineering

Rank	Country	Documents	Citable documents	% citable document	Citations	Self-citations	% self-citations	Citations per document	H index
1	South Africa	19163	18819	98.20	107903	19301	17.89	5.63	93
2	Algeria	13678	13566	99.18	58389	13228	22.65	4.27	69
3	Tunisia	12038	11924	99.05	52116	12671	24.31	4.33	63
4	Nigeria	5586	5544	99.25	16924	3737	22.08	3.03	48
5	Morocco	5536	5465	98.72	30741	6337	20.61	5.55	57
6	Libya	956	950	99.37	3362	153	4.55	3.52	29
7	Ghana	789	774	98.10	2973	445	14.97	3.77	26
8	Cameroon	680	677	99.56	3233	771	23.85	4.75	25
9	Sudan	642	638	99.38	1502	151	10.05	2.34	21
10	Kenya	574	566	98.61	3249	239	7.36	5.66	29

Source: SCImago, Author analysis, 2017

Research in the field of Material Science is also important to a nation's technological development. It connects with the industries in terms of quality of material resources needed for production. Besides, the engineering field also relates with this field as a form of support for the production of technology-oriented outputs needed as inputs in other sectors of the economy. Table 7 shows South Africa still topping the chart in Africa having produced a total of 10,956 documents. Algeria, Tunisia and Morocco are next ranked to South Africa. Notably in this field is the introduction of Cote d'Ivoire and Senegal to the table for the first time even though they occupy the bottom position in the ranks. West African countries are more engaged in research in this field of science. In terms of self-citation, the North African countries recorded the higher percentage in this analysis.

Table 7. Top 10 most productive countries in Africa in Material Science

Rank	Country	Documents	Citable documents	% citable documents	Citations	Self-citations	% self-citations	Citations per document	H index
1	South Africa	10956	10816	98.72	107086	18712	17.47	9.77	84
2	Algeria	8254	8194	99.27	57403	13266	23.11	6.95	69
3	Tunisia	6944	6839	98.49	48937	13668	27.93	7.05	59
4	Morocco	5717	5675	99.27	48726	10354	21.25	8.52	77
5	Nigeria	2485	2462	99.07	16032	3608	22.50	6.45	56
6	Cameroon	466	463	99.36	2819	613	21.75	6.05	24
7	Libya	396	392	98.99	2524	160	6.34	6.37	27
8	Ghana	320	314	98.13	1655	159	9.61	5.17	19
9	Côte d'Ivoire	314	314	100.00	1204	147	12.21	3.83	16
10	Senegal	287	285	99.30	2397	423	17.65	8.35	26

Source: SCImago, Author analysis, 2017

Table 8. Top 10 most productive countries in Africa in Medicine

Rank	Country	Documents	Citable documents	% citable documents	Citations	Self-citations	% self citations	Citations per document	H index
1	South Africa	46656	40847	87.55	744980	136527	18.33	15.97	239
2	Nigeria	19456	18421	94.68	134820	28891	21.43	6.93	104
3	Tunisia	15890	14211	89.43	101060	14291	14.14	6.36	89
4	Morocco	11773	10179	86.46	55312	6723	12.15	4.7	79
5	Kenya	9828	9225	93.86	191588	31385	16.38	19.49	143
6	Uganda	6522	6100	93.53	124757	20499	16.43	19.13	119
7	Tanzania	5899	5638	95.58	104297	17546	16.82	17.68	106
8	Ethiopia	4763	4582	96.20	47508	9552	20.11	9.97	71
9	Ghana	4248	4016	94.54	65276	7966	12.20	15.37	90
10	Cameroon	3850	3631	94.31	46576	7607	16.33	12.1	75

Source: SCImago, Author analysis, 2017

In Africa, Medicine is a field that still needs improvement in terms of research and human capacity development. Africans are the most travelled for medical attention in the world presently, according to the statistics on Medical tourism. Mostly, the destination is to Asian countries especially India, and some other countries such as Saudi Arabia, Germany, Israel etc. It can be deduced that in the field of medicine, as shown in Table 8, aside South Africa, research into this field is relatively low in West African countries. Considering the quality of publication among authors from these countries in the region, South Africa, Kenya, Nigeria and Uganda received higher h-index over all other countries.

4 Conclusion

The purpose of this paper which is to assess research performance of African countries in selected fields of S&T with respect to seven subject areas has been undertaken and with revealing inferences. Relating this outcome to realizing the AU 2063 agenda by member countries, there is a ray of hope in its attainment, although more commitment in the area of research and funding is needed. A particular case is that of the Medicine field where most of the citizens of countries such as Nigeria and others still embark on medical tourism to Asia and other European countries. Although the case of South Africa is different from that of other African countries in this regard, the country has capacities and physical infrastructure to attend to medical issues of his citizens, hence record low figure in medical tourism. South Africa tops the chart of the most productive countries in Africa in all the S&T field and occupy a position of 34th in the world. A closer look on the country next to South Africa, which is Nigeria, occupies 52nd position in the world. It can be deduced from the outcome that countries such as South Africa, including some North African countries like Morocco, Tunisia, Algeria, etc., enjoy adequate funding and maintain a clear strategic direction towards aligning their national developmental priorities to their research orientation. Besides, they have been able to structure and functionalize their National Innovation Systems (NIS) such that industrial needs inform their research priorities and knowledge acquisition.

In conclusion, the overall performance of African countries as it concerns this paper is promising and could be said to align towards realizing the regional goal. However, there is need for more coordinated and collaborative effort across the regions where it seems to be more productive. To this end, intra-African collaboration that is geared towards promoting knowledge development between researchers from low and high performing countries in Africa should be encouraged.

References

1. United Nations Economic Commission for Africa (UNECA): Africa's science, technology and innovation policies-national, regional and continental. In: Assessing Regional Integration in Africa (ARIA VII): Innovation, Competitiveness and Regional Integration, Chap. 5, pp. 83–104 (2016)
2. United Nations Economic Commission for Africa (UNECA): African Science Technology and Innovation Review 2013. Economic Commission for Africa, Ethiopia (2014)

3. Chaman, S.M., Dharani, K.P., Biradar, B.S.: Mapping of chemical science research in India during 2005–2014. Int. J. Inf. Dissemination Technol. **7**(1), 71–73 (2017)
4. Glanzel, W.: Bibliometrics as a research field: a course on theory and application of bibliometric indicators. Course Handout (2003). http://nsdl.niscair.res.in/jspui/handle/123456789/968. Last accessed 18 May 2017
5. Pouris, A.: Is Scientometrics in a crisis? Scientometrics **30**, 397–399 (1994)
6. Mingers, J., Leydersdorff, L.: A review of theory and practice in Scientometrics. Eur. J. Oper. Res. **246**, 1–19 (2015)
7. Hiremath, R., Gourikeremath, G., Hadagali, G., Kumbar, B.D.: India's Science and Technology output, 1989–2014: a scientometric analysis. Libr. Philos. Pract. (e-journal). Paper 1367 (2016)
8. Gupta, B.M., Bala, A., Kshitig, A.: S&T publications output of India: a scientometric analyses of publications output, 1996–2011. Libr. Philos. Pract. (e-journal). Paper 921 (2013)

Author Index

Printed in the United States
By Bookmasters

Printed in the United States
By Bookmasters